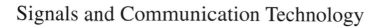

Signals and Communication Technology

For further volumes:
http://www.springer.com/series/4748

Hirokazu Yamanoue · Masaki Emoto
Yuji Nojiri

Stereoscopic HDTV

Research at NHK Science and Technology Research Laboratories

 Springer

Dr. Hirokazu Yamanoue
Patents Division
Science and Technology Research
    Laboratories
Japan Broadcasting Corporation (NHK)
1-10-11 Kinuta, Setagaya-ku
Tokyo 157-8510, Japan

Dr. Yuji Nojiri
Multimedia & Video Information Systems
    Division
NHK Integrated Technology Inc.
1-4-1 Jinnan, Shibuya-ku
Tokyo 150-0041, Japan

Dr. Masaki Emoto
Human & Information Science Division
Science and Technology Research
    Laboratories
Japan Broadcasting Corporation (NHK)
1-10-11 Kinuta, Setagaya-ku
Tokyo 157-8510, Japan

ISSN 1860-4862
ISBN 978-4-431-54022-9   ISBN 978-4-431-54023-6 (eBook)
DOI 10.1007/978-4-431-54023-6
Springer Tokyo Heidelberg New York Dordrecht London

Library of Congress Control Number: 2012942792

Printed on acid-free paper

Springer is part of Springer Science+Business Media (www.springer.com)

Contents

Chapter 1
Individual Differences in 3-D Visual Functions

Abstract Regarded from the perspective of human visual characteristics, 3-D television differs from conventional flat TV in that the viewer employs 3-D visual functions to perceive the image depth. Various measurements of these 3-D visual functions have been made to investigate the basic characteristics of the visual functions. It is now more important than before to grasp the characteristics of the visual functions to view 3-D images in order to further develop this field. The abstracts, based mainly on clinical ophthalmological tests of 3-D visual functions, are outlined below. Additionally, in anticipation of future 3-D television broadcasting services provided for large, various audiences, it is essential to grasp individual differences and variability in visual functions. For example, human visual system has limits to fuse left and right images with binocular parallax, and when the binocular parallax was greater than the limits, diplopia (double vision) is perceived. At the same time, depth perception also fails and the result is extremely unpleasant for the viewer. It is therefore necessary to grasp the limits to the binocular parallax within which left and right images can be fused by binocular vision. The individual differences of the 3D visual functions are very large, and this makes it important to grasp individual differences in 3-D visual functions across a large sample of people. The results for various 3-D visual functions are presented here for large populations.

Keywords Binocular parallax • Binocular vision • Dynamic stereopsis • Fusion • Stereoacuity • Stereopsis

H. Yamanoue et al., *Stereoscopic HDTV: Research at NHK Science and Technology Research Laboratories*, Signals and Communication Technology, DOI 10.1007/978-4-431-54023-6_1, © Springer Japan 2012

1.1 Stereopsis

1.1.1 Visual Acuity and Stereoacuity

The Landolt broken ring is used as the standard tool for measuring visual acuity. The score is the reciprocal value of the size of the smallest break that can be discerned by the eye. The ability to distinguish a gap of 1 arc minute is given the decimal visual acuity value of 1.0; distinguishing a gap of 0.5 arc minutes gives a decimal visual acuity value of 2.0. Various other tests of visual acuity do also exist, and among them the test of the ability to align two line segments in fact obtains finer results than are possible with Landolt's broken rings. This is known as vernier acuity and it reaches to approximately 2 arc seconds. Threshold of depth perception, or stereopsis is generally measured by dicho-optical methods such as polarizing filter to show left image to only left eye, and right image only to right eye. Well-known commercially-available tests include the Titmus circle test. These tests make it possible to test stereoacuity in steps for parallaxes ranging from 800 to 40 arc seconds.

1.1.2 Ocular Position

The term ocular position is employed in both monocular and binocular viewings but here we are referring to the binocular ocular position, which have great significance in relation to stereopsis. We call the ocular position used for normal binocular vision the functional binocular position. The problem is relative misalignment between left and right eye from the normal functional binocular position. This is referred to as strabismus and those who have strabismus in their infancy often have imperfect stereopsis. On the other hand, phoria is referred to as the ocular position where the relative misalignment between left and right eye is not observed when two images are fused, but the misalignment is evident when the fused stimuli are removed, while stereopsis generally functions normally. Strabismus and phoria may be referred to as horizontal, vertical, or cyclophoria, depending on the direction of misalignment. Ocular position tests are performed to discover these positions of misalignment.

1.1.3 Simultaneous Perception, Fusion, and Stereopsis

Worth et al. showed normal binocular vision as a hierarchical processes [1]:

1. Simultaneous perception
   This is the function for watching an object simultaneously with two eyes. The slightly different images entering each eye are perceived simultaneously.

2. Fusion

This is the function by which images received separately by the left and right eyes are perceived as a single image. The images received simultaneously by the two eyes are perceived as one image.

3. Stereopsis

This is the function in binocular vision by which the left and right images are resolved with a sense of depth.

These hierarchical processes do not function separately, but each of these elements is essential for the achievement of normal binocular 3-D vision. Accordingly, each of these hierarchies are examined separately in functional testing of binocular vision.

1.1.4 Fusional Range

The spatial zone two retinal images will be fused into a single image without eye movement is known as Panum's area [2]. Outside this area, eye movement (i.e. convergence and divergence) can be evoked to obtain binocular fusion. This eye movement is evoked by the binocular retinal disparity which is the difference in location between left and right corresponding points on retinae. Once the retinal disparity is detected, the two eyes move in the direction that will decrease the retinal disparity. These eye movements of convergence and divergence are known as fusional vergence. In binocular stereopsis, the fusional limits are exist both convergence and divergence side. The range between the two fusional limit is known as the fusional range. If the binocular parallax increases while the left and right images are being perceived as a single image and those limits are reached, then double vision will result (break point). This is known as the fusional vergence limit. Conversely, if the binocular parallax diminishes while the left and right images are being perceived with double vision, the two images do not fuse at the fusional vergence limit obtained when increasing the disparity. Fusion then happens only at a smaller parallax (recovery point). In effect, the fusional vergence limit exhibits hysteresis.

1.1.5 Dynamic Stereopsis

The binocular 3-D vision testing described above uses still images. The use of moving images makes it possible to test the feasibility of stereopsis with regard to temporal changes of shape and binocular parallax. Temporal changes come up against the limitation of the time it takes to achieve stereopsis, so the requirements here would appear to be stiffer than viewing still images. One does, however, occasionally find cases who cannot perceive depth with still images but can perceive depth with moving images.

Frequency (by individual)

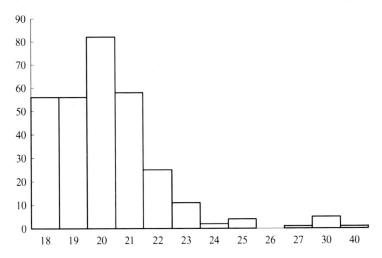

Fig. 1.1 Age distribution

1.1.6 Aniseikonia

Corrections by different lens power is required to those who have unbalanced refraction between left and right eyes (anisometropia) result in the different in retinal image size between left and right retinae. Lenses of very different powers must be used when the anisometropia are severe, then the difference in retinal image size is too large for the viewer to fuse them easily. This condition is known as aniseikonia. Visual tests for the aniseikonia is performed by delivering images of the same size separately to each eye to determine whether size differences are perceived. The degree of the aniseikonia is found by delivering images of different sizes to each eye which ones are perceived as having the same size.

1.2   An Example of Measurement of Visual Function

The following is an illustration of the application of measuring visual function. Given that the method of measurement adopted, depending on the item being measured, differs from that of clinical ophthalmology, a direct comparison of the results cannot be made directly. However, they sufficiently display the diversity of visual functions on a person-by-person basis. While data collection was unavailable for all individuals depending on the measured item, the target population for this measurement included 179 male and 122 female subjects. Ages were weighted toward university students, with an average age in the dataset of 20.42 years, a standard deviation of 3.62 years, and a range of 18–67 years. Figure 1.1 illustrates the age

Frequency (by individual)

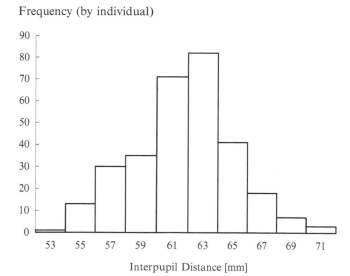

Fig. 1.2 Distance between pupils

distribution. One hundred and thirty-eight subjects required no visual corrections, 159 corrected for myopia, 1 corrected for hyperopia, and 3 unknown.

1.2.1   Interpupil Distance

Using a pupilometer (Hoya Digital Pupilometer RC-810; Tokyo, Japan), interpupil distance was measured when at a fixation distance of 0.6 m. The smallest unit of measurement was 0.5 cm. Figure 1.2 illustrates the distribution of the 301 subjects. There was a marginal significance of $p=0.083$ with respect to goodness of fit to a normal distribution.

1.2.2   Stereoacuity

Using a stereopsis screening chart, the threshold stereoacuity at a visual distance of 0.4 m was measured.

Figure 1.3 illustrates the distribution of stereoacuity for 295 subjects. Randot stereotests (displayed parallax: 20, 25, 30, 40, 50, 70, 100, 140, 200, 400 arc seconds; stereo optical) were used on 284 subjects, and Titmus circle tests (displayed parallax: 40, 50, 60, 80, 100, 140, 200, 400, 800 arc seconds; stereo optical) were used on 11 subjects, in order to combine the two sets of data, consolidation of

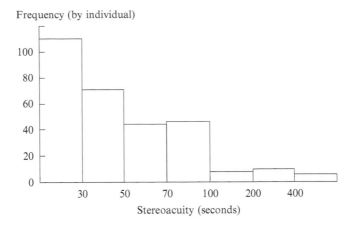

Fig. 1.3 Stereoacuity

Fig. 1.4 Apparatus

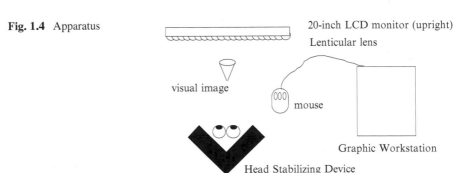

horizontal intervals was conducted at binocular parallaxes of 0–30, 30–50, 50–70, 70–100, 100–200, and 200–400 arc seconds. For example, subjects included in the 30–50 interval were those who were unable to perceive depth under the Randot stereotest at 30-arc second targets who were able to do so at 40 arc seconds, or those who were able to perceive depth of all targets in the Titmus circle test.

1.2.3 Ocular Position

The apparatus illustrated in Fig. 1.4 was used to measure misalignment of eye (or ocular) position and fusional limit. A graphic workstation was used as a signal source for both the presentation of test images and acquisition of subject responses. In order to present separate images to each eye of the subject, a 20-inch LCD monitor fitted with a lenticular lens was placed upright. In doing so, two images were presented side by side on the monitor, thereby presenting to the subjects' left and right eyes separately.

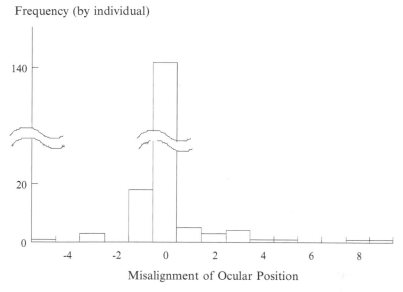

Fig. 1.5 Distribution of misalignment of the ocular position

1.2.4   Misalignment of Ocular Position

A perpendicular line was displayed to the left eye, and a small circle was displayed to the right eye, this produced a state whereby the subject had interrupted fusion within the display frame. The subject adjusted the position of the small circle such that both images appeared to be overlapping each other. The perpendicular line and the small circle would overlap if there were no misalignment of the ocular position; however, in the case of ocular misalignment the position of the line and the circle would misalign reflected subject's ocular misalignment. The misalignment of ocular position was measured with respect to the degree to which the images were misaligned. Figure 1.5 illustrates the data distribution of 180 individuals.

In comparison to the distribution measured by clinical ophthalmologists [3] these measurement results were concentrated in high amounts at an ocular position misalignment of zero. The method of measurement referred to in the literature is not clear; however, clinical ophthalmology in general makes use of large-size amblyoscopes and prisms, and because measurements are conducted by manipulating the full field of view it can be surmised that the such full field of view measurement method was used. Given that the current apparatus was used for stereoscopic television whereby a picture width (or frame) exists, fusion cannot be stated to have been entirely interrupted, and where manipulation and measurement were conducted only on the range of field of view within the picture width (or frame), it can be thought that the differing measurement methodologies may account for some differences.

Given that a 2–3° horizontal heterophoria is considered a physiological variance, the results of this measurement of eye misalignment yielded only 15 subjects with misalignments of ocular position greater than 2°, those subjects were selected from the subject group and regression analysis of the mean value of eye misalignment and fusional limit was conducted. The result of this analysis was a statistical probability of $p = 0.058$, implying that for those subjects with an eye misalignment of above 2°, the mean value of fusional convergence tended to be shifting in the direction of eye misalignment.

1.2.5   Fusional Limit

In measuring fusional limits, binocular parallax of a background image was unchanged at zero, but when a circular image having a radial pattern inside was overlapped, the binocular parallax changed in a stepwise manner.

The internal radial pattern also had binocular parallax and, if fused, was perceived as cone-shaped (as illustrated in Fig. 1.4). The cone shape is initially perceived as lying on a single flat surface together with the background image, however if fusion is possible after a step change in the quantity of binocular parallax, the cone shape is either perceived to jump out from the background image toward the viewer, or its backward motion is perceived to fade into the background. In cases of a large change in the binocular parallax whereby presented left and right images cannot be fused, a double image is perceived. This type of step-change of binocular parallax is consistent with sudden discontinuous changes in binocular parallax occurring in instances such as a scene change when viewing a stereoscopic television, and represents the strictest condition of measurement. In the measurement of fusional range within clinical ophthalmology, it is general for repetitive binocular parallax change to be measured; however, in such cases it is known that a subject's possible fusion range becomes even wider.

With respect to our 301 subjects, Fig. 1.6 illustrates the frequency distribution or histogram (left vertical axis) and cumulative relative frequency (right vertical axis) in cases where the change in the binocular parallax of the divergent side is greater than that of the display screen. The mean value of these data was 3.40° with a 1.59° standard deviation.

Figure 1.7 illustrates the frequency distribution or histogram (left vertical axis) and cumulative relative frequency (right vertical axis) in cases where the change in the binocular parallax of the convergent side is greater than that of the display screen. The mean value of these data was 5.04° with a 5.06° standard deviation.

In conclusion, the analysis above illustrates the variation and diversity of binocular visual function from person to person which come into play when viewing a stereoscopic television. It also underscores the need for a system design that takes into account such diversity in binocular visual function, where a fusionable binocular parallax image for one individual cannot be fused by another.

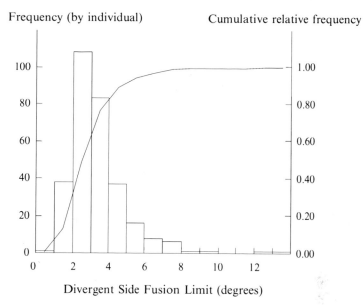

Fig. 1.6 Distribution of divergent side fusion limit

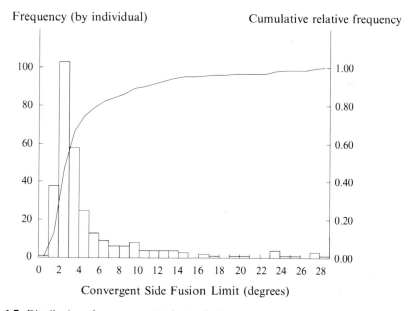

Fig. 1.7 Distribution of convergent side fusion limit

References

1. Worth, C. "Squint: Its causes, Pathology, and treatment", Philadelphia, Blakiston (1908)
2. Panum, P.L. "Physiologische Untersuchungen über das Sehen mit zwei Augen. Kiel: Schwerssche Buchhandl., (1858)
3. Ando, Tei Nakamichi, Yabe. "Study of Cyclic Movement in Nervous Asthenopia", Clinical Ophthalmology, 27, 7, pp. 871-880 (1973)

Chapter 2
Research on 3-D Image Distortions Caused by Recording and Viewing Conditions

Abstract This chapter reviews the studies of the distortion characteristic of stereoscopic images seen with shooting and viewing conditions needed in order to generate stereoscopic images which viewers would find natural and easy to watch. We start with the theoretical analysis of the basic features of stereoscopic images attributable to recording and viewing conditions, with special attention to the positioning of optical axes. The impact of the optical axes on the puppet-theater and cardboard effects was assessed by subjective evaluation. There were also subjective evaluations of distortion free conditions obtained from these theoretical studies with reference to the sizes of 3-D objects, the distances to them, and the naturalness of the stereoscopic effects. The comprehension of these distortion free conditions is fundamental to our understanding of the features of stereoscopic images.

Keywords 3-D image • Binocular parallax • Camera separation • Cardboard effect • Orthostereoscopic condition • Orthostereoscopy • Parallel camera configuration • Puppet theater effect • Recording condition • Shooting condition • Stereoscopic image • Toed-in camera configuration • Viewing condition

2.1 Recording Conditions for Stereoscopic Images

2.1.1 Introduction

Images captured by two cameras corresponding to the left and right eyes arrive as a double image according to the distance between the lenses. This shift between the two images is known as binocular parallax and it is also a key factor in depth perception. The perception of depth and other stereoscopic effects can be produced by separating the left and right images by means of polarized glasses and projecting the

H. Yamanoue et al., *Stereoscopic HDTV: Research at NHK Science and Technology Research Laboratories*, Signals and Communication Technology, DOI 10.1007/978-4-431-54023-6_2, © Springer Japan 2012

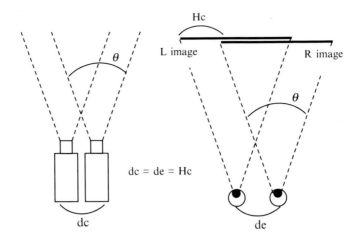

Fig. 2.1 An example of parallel camera configuration (distortion free conditions)

left camera image on the left eye and the right camera image on the right eye. The fact that two cameras are involved means that the range of recording conditions which has to be considered is greater than in the case of two-dimensional images. These conditions produce significant effects on the screen [1–5].

2.1.2   Setting the Optical Axes

There are two ways to set optical axes: in parallel or cross-eyed (toed-in). When the optical axes are parallel to each other (hereinafter referred to as the "parallel camera configuration"), the basic approach is to generate a horizontal shift between the left and right images during image acquisition or viewing in order to observe the object, which is at an infinitely distant point, at an infinite distance. If, as shown in Fig. 2.1, both the distance between the two cameras (dc) and the horizontal shift between the left and right images (Hc) are the same as the distance between the observer's left and right irises (de), and the viewing angle of the lenses corresponds to that of the screen (θ), the entire shooting space should, in theory, be visible in undistorted stereoscopic images. These conditions, details of which will be described in later chapters, set standards for conversion from the shooting space to the space of the stereoscopic image but it is difficult in broadcasting to satisfy these conditions in every case, where the existence of diverse recording and viewing conditions has to be assumed. It is, therefore, most important to analyze the situations in which these conditions are not met.

Figure 2.2 shows recording with crossed optical axes (hereinafter referred to as the "toed-in camera configuration"). The object is here located at the point where the optical axes converge and reproduced on the screen in this manner as a

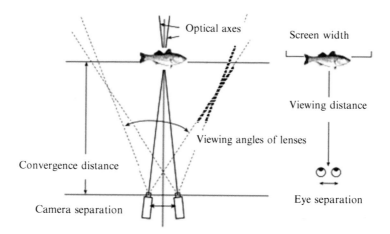

Fig. 2.2 An example of toed-in camera configuration

stereoscopic image. The intensity of the stereoscopic effect is influenced by the distance between the cameras, with variance around the point where the optical axes converge. This makes it possible to change where the object is reproduced in the stereoscopic image space, either closer to or further away from the viewer, by moving this convergence point back and forth. At the same time, the separation of the cameras can be altered to adjust the intensity of the stereoscopic effects. This facility in manipulating stereoscopic effects has often made toed-in camera configuration the method of choice in television program production, but it is susceptible to such visual distortions as the puppet-theater and cardboard effects with 3-D pictures. The next chapter explains the conditions that give rise to these visual distortions.

2.1.3   Camera Separation

The shift between the left and right images due to the use of separate cameras produces binocular parallax, which is a very important factor in depth perception in human vision. Camera separation, therefore, has a large impact on both the stereoscopic effects and depth perception. The recording distance to the object and the size of the subject are the most critical factors when adjusting camera separation, but consideration must also be given to the setting of the optical axes, focal length of lenses, viewing distance, and screen size. Generally, the cameras are set closer together if the subject is small and close to the camera, and further apart if the subject is large and distant from the camera. When recording the hatching of Japanese rice-fish (*medaka*) eggs for TV broadcast, the cameras were set 1 cm apart. When recording images of the earth from the space shuttle, the camera separation was 40–50 km.

2.1.4   Focal Length of the Lens (Viewing Angle of the Lens)

Changing the focal length of the lens typically changes the perspective. A shorter focal length makes foreground objects seem larger and background objects smaller, emphasizing visual depth. Whereas the shift caused by changing the distance between two cameras provides depth information through both eyes, perspective provides depth information through only a single eye.

2.2   The Basic Properties of Reproduced Space

2.2.1   Introduction

The determination of the appropriate recording and viewing conditions for stereoscopic images is a choice about how to convert real space into the space of the stereoscopic image. This is one of the most fundamental tasks in constructing a stereoscopic imaging system. Much effort has been made to analyze this conversion using the shift between the left and right images (binocular parallax) as a parameter. In particular, the conditions of theoretically reflecting recording space in stereoscopic image space (called "the orthostereoscopic condition") has been studied from various angles, including comparisons of subjective assessments with objective experimental results. The orthostereoscopic condition is fundamental to understanding the features of stereoscopic images [6, 7]. As this condition specifies very particular recording and viewing conditions, however, it is impossible to satisfy them all of the time in actual broadcasting. It is essential, therefore, to understand the basic features of stereoscopic images aside from the orthostereoscopic condition. We have made a particular study of the setting of optical axes, because the different ways of placing these axes create not only differences in the stereoscopic images formed but also made significant differences in the size and weight of the 3-D cameras. As the demand for efficient and effective program production increases, we need to obtain a precise understanding of the characteristics of these ways of placing optical axes in order to use them flexibly.

2.2.2   Models of Recording and Display Systems

In this section, we analyze how the recording space (real space) is projected on the viewing space (stereoscopic image space) under various recoding and viewing conditions, such as how the optical axes are set, camera separation, the focal length of lens, screen size, and viewing distance, using the models of recording and display systems shown in Figs. 2.3 and 2.4 and parameters shown in Table 2.1. For simplicity, we use the center of an image free of keystone distortion.

Fig. 2.3 Model of recording
system

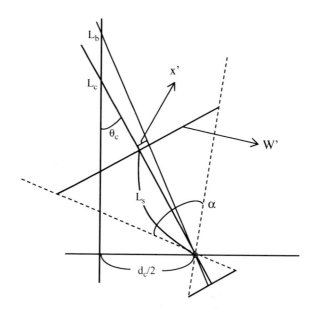

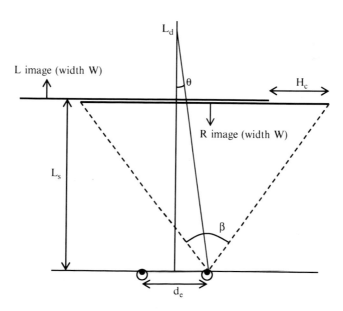

Fig. 2.4 Model of display system

Table 2.1 Parameters of recording and display models

d_c	Camera separation
d_e	Eye separation
L_b	Shooting distance
L_c	Convergence distance
L_s	Viewing distance
L_d	Position of a stereoscopic object
α	Viewing angle of lens
β	Viewing angle
θ_c	Camera convergence angle
H_c	Horizontal gap between L and R images (added at time of display)
W	Width of screen
W'	Width of virtual screen at the viewing distance of the recording model

The following equation can be established from Fig. 2.3:

$$(L_b - L_c) \times \sin\theta_c : x' = \left\{ (L_b - L_c) \times \cos\theta_c + \sqrt{\left(\frac{d_c}{2}\right)^2 + L_c^2} \right\} : L_s \qquad (2.1)$$

Assuming that θ_c is sufficiently small and

$$\sin\theta_c \cong \frac{d_c}{2L_c} \quad \cos\theta_c \cong 1,$$

we obtain

$$x' = \frac{(L_b - L_c) \cdot d_c \cdot L_s}{2L_b \cdot L_c} \qquad (2.2)$$

From Figs. 2.3 and 2.4, we obtain

$$x = \frac{W}{W'} \cdot x' + \frac{H_c}{2} \qquad (2.3)$$

For an eye convergence angle of 2θ when seeing a stereoscopic image, we derive

$$\tan\theta = \frac{d_e}{2L_d} = \frac{1}{L_s}\left(\frac{d_e}{2} - x\right) \qquad (2.4)$$

where we define the camera separation ratio as

$$\frac{d_c}{d_e} = a_1 \qquad (2.5)$$

and the magnification of an image on the retina as

$$\frac{W}{W'} = \frac{\tan \dfrac{\beta}{2}}{\tan \dfrac{\alpha}{2}} = a_2 \tag{2.6}$$

By substituting (2), (3), (5) and (6) into (4), we now have the final location L_d, where the stereoscopic image is formed, as follows:

$$L_d = \frac{1}{\dfrac{1}{L_s} - \dfrac{a_1 \cdot a_2}{L_c} + \dfrac{a_1 \cdot a_2}{L_b} - \dfrac{H_c}{L_s \cdot d_e}} \tag{2.7}$$

2.2.3   Parallel Camera Configuration

We assume in (7) that the distance to the convergence point is infinite. Regarding H_c, the horizontal shift of left–right images, we further assume that $H_c = d_e$, the infinite point during recording, will also be located at an infinite point during viewing. With these assumptions, (7) can be simplified to

$$L_d = \frac{L_b}{a_1 \cdot a_2} \tag{2.8}$$

As this equation shows, if the recording is performed by the parallel camera configuration and the amount of left–right horizontal shift is equal to the distance between the observer's irises $H_c = d_e$, then linearity can be retained between the shooting distance to the object and the distance at which the stereoscopic image is formed. An example is shown in Fig. 2.5. This shows the relationship between the shooting distance (L_b) and the distance to the 3-D images (L_d) in the parallel camera configuration under the recoding and viewing conditions shown in Table 2.2.

When an object is shot by the parallel camera configuration, the projected 3-D image moves toward the viewer as the distance between the cameras widens, but away from the viewer as the lens view angle expands.

2.2.4   Toed-In Camera Configuration

Unlike the parallel camera configuration, this method basically moves the convergence point back and forth without creating a horizontal shift between the left and right images.

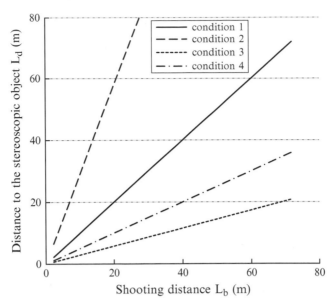

Fig. 2.5 Relationship between the shooting distance and the distance to 3-D objects shot in the parallel camera configuration

Table 2.2 Example recording and viewing conditions

Conditions	1	2	3	4	5	6	7	8
Camera configuration	P	P	P	P	T	T	T	T
Camera separation d_c (mm)	65	30	90	130	65	65	32.5	130
Viewing angle of lens α (°)	33.4	43.6	13.7	33.4	33.4	43.6	43.6	43.6
Convergence distance L_c (m)	∞	∞	∞	∞	4.5	4	1.8	6.1
Viewing angle β (°)	33.4	33.4	33.4	33.4	33.4	33.4	33.4	33.4
Viewing distance L_s (m)	4.5	4.5	4.5	4.5	4.5	4.5	4.5	4.5
Horizontal shift H_c (mm)	65	65	65	65	0	0	0	0

Note: *P* parallel, *T* toed-in

Assuming that $H_c = 0$ in (7), we obtain

$$L_d = \cfrac{1}{\cfrac{1}{L_s} - \cfrac{a_1 \cdot a_2}{L_c} + \cfrac{a_1 \cdot a_2}{L_b}} \qquad (2.9)$$

Figure 2.6 describes the relationship between the shooting distance (L_b) and the distance to 3-D images (L_d) shot by the toed-in camera configuration under the recoding and viewing conditions shown in Table 2.2. In the toed-in camera configuration, as we see in Fig. 2.6 and (9), L_d begins to undergo a major change when $L_c = a_1 \cdot a_2 \cdot L_s$. When $L_c < a_1 \cdot a_2 \cdot L_s$, L_d cannot be calculated by (9) if

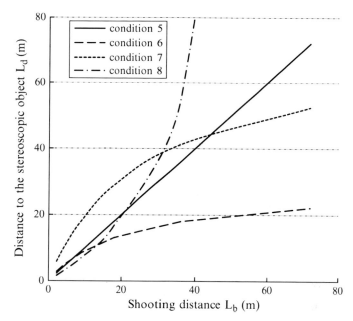

Fig. 2.6 Relationship between the shooting distance and the distance to 3-D objects shot by the toed-in camera configuration

the shooting distance L_b is large. This is a situation in which the eyes do not converge (or diverge) as the distance between the left–right irises changes. This situation does not produce immediate fusion. If $L_c = a_1 \cdot a_2 \cdot L_s$ [condition (5)], (9) is the same as (8), meaning that there is no distortion at the center of the screen, but that keystone distortion occurs in the peripheral area.

2.2.5   Summary

We have analyzed geometrical conversion from real space into stereoscopic images under various recording and viewing conditions. These studies focused on how optical axes are set, an issue which has received little attention in the past, and clarified the relationship between the shooting distance to the object and display position in 3-D image space.

The results show that this relationship remains linear when objects are shot using the parallel camera configuration. The relationship does, however, become radically nonlinear depending on the parameters if the objects are shot using the toed-in camera configuration. These analytical results will be referred to in later sections in relation to theoretical analyses of puppet-theater and cardboard effects.

2.3   Puppet Theater Effect

2.3.1   Introduction

Here, we make a theoretical analysis of how the choice setting method for the optical axes affects the puppet theater effect, and examine general images in terms of subjective evaluation.

2.3.2   Definition of the Puppet Theater Effect

The puppet theater effect is a size distortion characteristic of 3-D images [8–12]. It is well known that 3-D images of a shot target tend to look unnaturally small. How observers perceive the apparent size of an object in the image is also heavily influenced by such factors as their familiarity with the object and past experience. The puppet theater effect is not perceived as a physically measurable amount, therefore, but rather as a phenomenon to be grasped through subjective evaluation. In this paper, we evaluate the puppet theater effect quantitatively by defining the phenomenon as described below, assuming that the observers know the shot target fairly well.

MacAdam [13] discussed the distortion of size that occurs between an object in the foreground and one in the background with special attention to the relationship between the depth information (perspective) from 2-D images and depth informa-tion (binocular parallax) from 3-D images. He showed how the apparent size of the reproduced object can vary depending on the shooting distance and noted a possible link with the puppet theater effect. We follow his approach in this paper, regarding the puppet theater effect as a size distortion caused by the fact that the image's apparent size differs according to the shooting distance. Here, we calculate geo-metrically the relative size differences that arise between objects shot inside the image space due to disparities in the magnification of the object and its foreground and background. The results of this calculation provide standards for assessing the puppet theater effect. The absolute size may also contribute to this visual distortion, but this point is not discussed here.

Distortion of relative size can, of course, be perceived in either one of two ways; the object may look larger or smaller than it is in reality. The result depends on whether the object is in the background or foreground, the nature of the object, its patterns, and various other factors besides. These patterns and other characteristics do affect whether a background object looks smaller or larger when the foreground serves as the reference point and whether a foreground object looks smaller or larger when the background serves as the reference point. This report does not address these issues. Rather, we examine the puppet theater effect as a size distortion result-ing from dimensional inconsistency between the targets shot.

Starting from these assumptions, we here describe how the puppet theater effect is likely to occur when the image's apparent size is dependent on the shooting distance. If the shooting distance to objects in the background or foreground is known, the appropriate magnification of their images can also be calculated and the ratio of their respective magnifications can be used as a predictive reference for the puppet theater effect. The relationship between this objective measure and subjective evaluation will be discussed in Sect. 2.3.5. If W_b is the real size in the shooting space (real space) of the object and W_r is its apparent size in the stereoscopic image space, we obtain

$$W_r = \frac{L_d}{L_b} \cdot \frac{\tan\dfrac{\beta}{2}}{\tan\dfrac{\alpha}{2}} \cdot W_b = \frac{L_d}{L_b} \cdot a_2 \cdot W_b \tag{2.10}$$

where the magnification of image W_r / W_b is commonly known as the lateral magnification of the lens's optical system.

2.3.3   *Geometrical Analysis 1 (Optical Axes in Parallel)*

From (8) and (10), the magnification of the image (lateral magnification) $Ms \left[\equiv \dfrac{W_r}{W_b} \right]$

can be expressed as

$$Ms = \frac{W_r}{W_b} = \frac{1}{a_1} \tag{2.11}$$

This indicates that, with the parallel camera configuration, the magnification of the image is determined only by the ratio of the camera separation to eye separation (camera interval ratio a_1). Using the definition of the puppet theater effect described above, however, the puppet theater effect does not occur here because shooting distance L_b is not included. Figure 2.7 shows the relationship between Ms and L_b when 3-D images are shot by a parallel camera configuration under the conditions shown in Table 2.2.

2.3.4   *Geometrical Analysis 2 (Crossed Optical Axes)*

From (9) and (10), we obtain

$$Ms = \frac{W_r}{W_b} = \frac{1}{\dfrac{1}{L_s} - \dfrac{a_1 \cdot a_2}{L_c} + \dfrac{a_1 \cdot a_2}{L_b}} \cdot \frac{a_2}{L_b} \tag{2.12}$$

Figure 2.8 indicates the relationship between Ms and L_b when 3-D images are shot by the toed-in camera configuration under the conditions shown in Table 2.2.

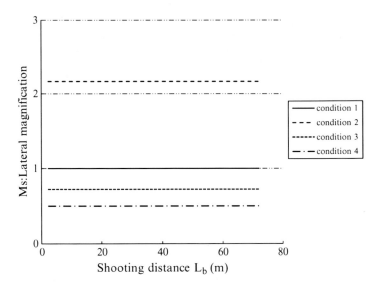

Fig. 2.7 Ratio of real size to apparent size of 3-D objects shot by parallel camera configuration (=Ms)

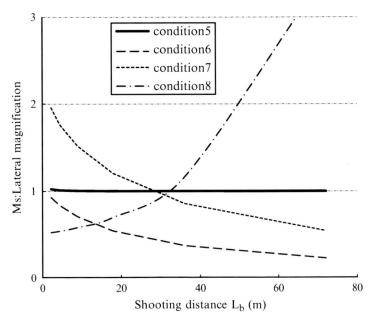

Fig. 2.8 Ratio of real size to apparent size of 3-D objects shot by toed-in camera configuration (=Ms)

If $L_c = a_1 \cdot a_2 \cdot L_s$, (12) and (11) are identical; if not, the magnification Ms is dependent on shooting distance L_b. If $L_c > a_1 \cdot a_2 \cdot L_s$, under conditions 6 and 7 in Fig. 2.8, the background object is shown smaller than the foreground object. If $L_c < a_1 \cdot a_2 \cdot L_s$, the foreground object is shown smaller than the background object. When $L_c \ll a_1 \cdot a_2 \cdot L_s$, the magnification of the object's apparent size becomes highly dependent on the shooting distance, making appearance of the puppet theater effect probable. As described above, perception of the effect is heavily dependent on the patterns and nature of the image. For quantitative referencing, information is needed on the shooting distances to the background and foreground objects.

2.3.5   Subjective Tests (Relationship with Binocular Parallax)

We earlier defined the puppet theater effect as a distortion of the relative sizes of objects shot in the image space. It follows that no puppet theater effect is going to occur with the parallel camera configuration because the magnification of the image is constant and independent of shooting distance. With the toed-in camera configuration, on the other hand, the magnification is sometimes dependent on shooting distance and this makes the puppet theater effect more likely.

This section shows the results of an experiment in which the size of an object shot by the toed-in camera configuration was evaluated subjectively on the assumption that the shooting distance, which could be either in the foreground or in the background, was known. The results are then compared with the predictive reference for the puppet theater effect as defined in this paper.

This predictive reference Ep for the puppet theater effect can be expressed by the following, where Ms_f is the magnification (lateral magnification) of the object that forms the foreground, and Ms_b the magnification (lateral magnification) of the object that forms the background:

$$Ep = \frac{Ms_f}{Ms_b} \qquad (2.13)$$

For subjective evaluation, we used nine different images obtained by shooting the object shown in Fig. 2.9 in three different ways (with different shooting distances and viewing angles of the lens and three different camera separations). Regardless of the recording conditions, a life-size 3-D image of the object was always formed on the screen (screen size: 120 inch diameter). Observers were asked to evaluate the size of the 3-D image on a scale of 1–5, as follows: (1) small, (2) rather small, (3) normal size, (4) rather large, and (5) large.

Table 2.3 shows the recording and viewing conditions and also Ep as calculated by (13). To calculate Ep, the mannequin at the center of the screen was used as the foreground; the door 4.5 m behind the mannequin in the hallway was used as the background. Figure 2.10 shows the relationship between the subjectively perceived size of the object in the foreground and predictive reference Ep for the puppet theater

Fig. 2.9 Object for subjective evaluation of the puppet theater effect

Table 2.3 Recording and viewing conditions

Conditions	1	2	3	4	5	6	7	8	9
Type	A	A	A	B	B	B	C	C	C
Viewing angle of lens α (°)	51.3	51.3	51.3	27.0	27.0	27.0	18.2	18.2	18.2
Distance to foreground object (m)	3.0	3.0	3.0	6.0	6.0	6.0	9.0	9.0	9.0
Distance to background object (m)	7.5	7.5	7.5	10.5	10.5	10.5	13.5	13.5	13.5
Convergence distance L_c (m)	3.0	3.0	3.0	6.0	6.0	6.0	9.0	9.0	9.0
Camera separation d_c (mm)	65	95	125	65	95	125	65	95	125
Viewing angle β (°)	33.4	33.4	33.4	33.4	33.4	33.4	33.4	33.4	33.4
Viewing distance L_s (m)	4.11	4.11	4.11	4.11	4.11	4.11	4.11	4.11	4.11
a_1	1.00	1.46	1.92	1.00	1.46	1.92	1.00	1.46	1.92
a_2	0.63	0.63	0.63	1.25	1.25	1.25	1.88	1.88	1.88
$L_c/(a_1 \cdot a_2 \cdot L_s)$	1.16	0.79	0.60	1.17	0.80	0.61	1.16	0.80	0.61
Ep	1.21	0.61	0.01	1.11	0.81	0.52	1.07	0.87	0.68

effect (as defined in this paper). Figure 2.10 reveals a high correlation coefficient of 0.957 between the subjective evaluation of the foreground object and predictive reference Ep.

This experiment was designed to compare the results of the geometrical calculations of relative sizes of the mannequin and the background with the subjective evaluation of the mannequin's size when a life-size image of the mannequin (shown in Fig. 2.9) was displayed on a fixed 120-inch screen. The patterns in this case tend to reduce the apparent size of the mannequin because the background fills a lot of the

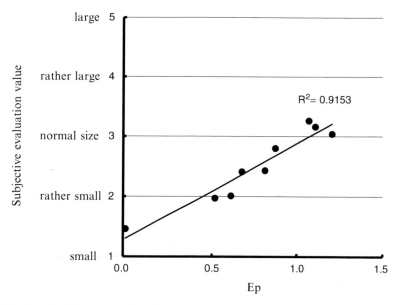

Fig. 2.10 Relationship between Ep and subjective evaluation

screen. As explained at the beginning, the puppet theater effect is here considered only in terms of the relative sizes of the objects. Many more images have to be examined, however, as it is sometimes difficult to determine whether it is the foreground or the background that is dominant in 3-D images, depending on the picture patterns.

2.3.6   Summary

We have studied the puppet theater effect theoretically in terms of how the optical axes are set and the conditions that give rise to it as one of the characteristic distortions that occur in stereoscopic images. First, we defined the puppet theater effect as a size distortion due to inconsistency between the relative sizes of objects within the image space. The puppet theater effect will not occur in the case of the parallel camera configuration because the magnification of the image is constant and independent of the shooting distance. In the case of the toed-in camera configuration, however, the puppet theater effect is likely occur because the magnification of the image is dependent on the shooting distance (except when $L_c = a_1 \cdot a_2 \cdot L_s$). Using photographed images, we studied the correlation between predictive reference Ep for the puppet theater effect based on theoretical considerations for binocular parallax and the subjective evaluations, and confirmed that Ep is consistent with those subjective impressions.

2.4   Cardboard Effect

2.4.1   Introduction

The cardboard effect is a phenomenon in which the scene looks layered, that is, consisting of a flat object and flat background, although the observer can distinguish the physical relationship between the two objects [14, 15]. This phenomenon needs to be addressed comprehensively as a shape perception issue. In this paper, however, as in the case of the puppet theater effect, we analyze the cardboard effect theoretically and then examine general images subjectively with binocular parallax serving as the only condition.

2.4.2   Geometrical Analysis of the Cardboard Effect

If depth ΔL_b at shooting distance L_b in real space is reproduced as depth ΔL_d in the stereoscopic image space, optical depth magnification Md can then be expressed as:

$$Md = \frac{\Delta L_d}{\Delta L_b} \tag{2.14}$$

Md: depth magnification.

   In order to reproduce the shape of an object precisely during conversion from shooting space to viewing space, magnification *Ms* (lateral magnification) for the apparent size must be the identical to depth magnification *Md* (axial magnification). Here, thickness *Ec*, which is the predictive reference for the cardboard effect, is defined as:

$$Ec = \frac{Md}{Ms} = \frac{\dfrac{dL_d}{dL_b}}{Ms} \tag{2.15}$$

Ec: thickness (predictive reference for the cardboard effect).

   The cardboard effect is likely to occur if the depth magnification is smaller than that of breadth in (15), which makes *Ec* small. The other factors likely to contribute to this phenomenon are shading, contour, texture, and motion parallax. This paper focuses on geometrical analysis based on binocular parallax.

   When shot with the parallel camera configuration, we obtain, from (8) and (11)

$$Ec = \frac{Md}{Ms} = \frac{\dfrac{1}{a_1 \cdot a_2}}{\dfrac{1}{a_1}} = \frac{1}{a_2} \tag{2.16}$$

Equation (16) indicates that the cardboard effect with the parallel camera configuration is dependent on the ratio of the viewing angle of the lens to the observer's viewing angle.

For shooting with the toed-in camera configuration, we obtain, from (9) and (12)

$$Ec = \frac{Md}{Ms} = \frac{\dfrac{dL_d}{dL_b}}{Ms} = \frac{a_1}{L_b} \cdot \frac{1}{\dfrac{1}{L_s} - \dfrac{a_1 \cdot a_2}{L_c} + \dfrac{a_1 \cdot a_2}{L_b}} \qquad (2.17)$$

If the image of the object forms near the screen, L_b equals L_c and (17) can be simplified to:

$$Ec = a_1 \cdot \frac{L_s}{L_c} \qquad (2.18)$$

This indicates that, in order to reduce the cardboard effect with the toed-in camera configuration, it is effective to increase the distance between the cameras, shorten the distance to the point of convergence, and increase the viewing distance.

2.4.3 Subjective Evaluation Test 1: Relationship with Binocular Parallax

In the previous section, we defined the degree of thickness Ec as the reference thickness of the cardboard effect. In this section, we used the still picture shown in Fig. 2.11, in order to obtain a better understanding of the relationship between *Ec* and the subjective data. This picture was presented as a 3-D image under the recording and viewing conditions shown in Table 2.4. The thickness of the 3-D image was evaluated on a scale of 1–5, as follows: (1) not thick at all, (2) not very thick, (3) somewhat thick, (4) thick, and (5) very thick.

The results are shown in Fig. 2.12, in which the subjective space (from 1 to 5 on the vertical axis) is standardized according to successive categories. *Ec* obtained from (18) is represented on the lateral axis. The coefficient of correlation is 0.93, which is satisfactory. In this experiment, only one object was shot and the lighting conditions were constant, in order to eliminate other factors (such as contours and shading) that contribute to the cardboard effect; we focused on the results obtained through geometrical analysis relying solely on binocular parallax.

2.4.4 Subjective Evaluation Test 2: Effects from a Background

The cardboard effect is influenced not only by binocular parallax, which provides depth, but by other factors as well, including shading, contours, the existence (or absence of) any background, and motion parallax. In this section, we discuss the tests we conducted on the effects of motion parallax and the existence of any background.

Fig. 2.11 Shooting target for
subjective evaluation tests on
the cardboard effect

Table 2.4 Recording and viewing conditions in the subjective evaluation tests on the cardboard effect

Conditions	1	2	3	4	5	6	7	8	9
Camera configuration	Toed-in								
Viewing angle of lens α (°)	43.6	43.6	43.6	13.7	13.7	13.7	5.7	5.7	5.7
Shooting distance (m)	3.3	3.3	3.3	11.0	11.0	11.0	26.4	26.4	26.4
Convergence distance (m)	3.3	3.3	3.3	11.0	11.0	11.0	26.4	26.4	26.4
Camera separation d_c (mm)	13	39	66	43	130	216	104	311	513
Viewing angle β (°)	43.6	43.6	43.6	43.6	43.6	43.6	43.6	43.6	43.6
Viewing distance L_s (m)	3.3	3.3	3.3	3.3	3.3	3.3	3.3	3.3	3.3
a_1		0.20	0.60	1.02	0.66	2.00	3.32	1.60	4.78	7.89
Ec		0.20	0.60	1.02	0.20	0.60	1.00	0.20	0.60	0.99

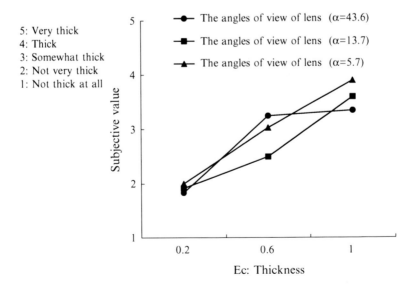

5: Very thick
4: Thick
3: Somewhat thick
2: Not very thick
1: Not thick at all

The angles of view of lens (α=43.6)
The angles of view of lens (α=13.7)
The angles of view of lens (α=5.7)

Fig. 2.12 Relation between subjective values and Ec

Fig. 2.13 Background with stereoscopic images

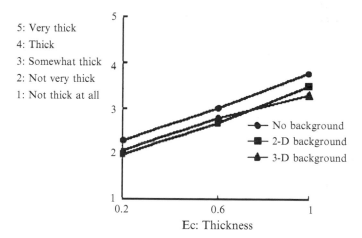

Fig. 2.14 Background effects

We initially set three criteria for gauging the effects of any background, adding them to the criteria mentioned in Table 2.4, viz. the absence of any background (see Fig. 2.11); the existence of a flat background in the rear of the screen with uniform parallax; and the existence of a stereoscopic background of successive parallaxes extending from near the front to the back of the screen. Figure 2.13 provides an example of the latter, in which the 3-D image appears in the left eye. Subjective evaluation tests were conducted on a scale of 1–5 in the same manner as the previous section, standardizing the subjective spaces by arranging the resulting data into successive categories.

Figure 2.14 shows the test results. We used variance analysis to produce averaged data for the factor of the viewing angle of the lens (factor of the focal length of the lens), because a main effect or interaction were not found in these factors.

Fig. 2.15 Motion parallax

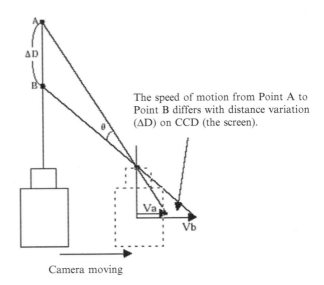

The speed of motion from Point A to Point B differs with distance variation (ΔD) on CCD (the screen).

Camera moving

As we see in the figure above, the cardboard effect seems more noticeable with than without a background. This finding corroborates reports that the cardboard effect occurs when there is discontinuous parallax between the background and the object. As the figure also shows, however, there is no significant difference regarding the two set background conditions, namely, the one with uniform parallax for the 3-D image to be evaluated and the other for various parallaxes from foreground to background. This suggests that the cardboard effect is influenced by the existence of background per se, regardless of the setting and background type.

2.4.5 Subjective Evaluation Test 3: Effects of Motion Parallax

Next, we discuss the influence of motion parallax. As Fig. 2.15 shows, motion parallax is expressed in terms of depth information, based on differences in the speed of movement of the object with reference to variation in the distance (ΔD). Here, we give further consideration to the situation where the same object as above is rotated slowly instead of moving the camera. In this case, the differences between the speeds of motion of the center and peripheral areas of the object on the screen provide the keys to depth information (see Fig. 2.16).

The experimental conditions, objects, and methods of evaluation are the same as in the earlier experiment (see Table 2.4). The tests include the full rotation of the object shown in Fig. 2.11 through 360° in 15 s at a constant speed. The amount of motion parallax in each set of conditions shown in Table 2.4 remains the same. It should be noted that there is no background image. The results are shown in Fig. 2.17. As in Fig. 2.14, we produce averaged data for the factor of the viewing angle of the lens because neither a main effect nor interactions were found.

Fig. 2.16 Motion parallax
when the object rotates

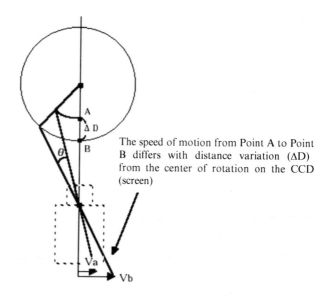

The speed of motion from Point A to Point B differs with distance variation (ΔD) from the center of rotation on the CCD (screen)

Fig. 2.17 Effects of
motion parallax

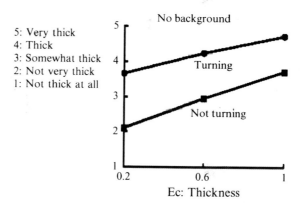

5: Very thick
4: Thick
3: Somewhat thick
2: Not very thick
1: Not thick at all

As this figure indicates, there is a significant difference in evaluations depending on whether the object is turning. The subjective perception of thickness is strong when the object is turning. Even when thickness Ec is low, the finding increases substantially; this is because the motion parallax applied to the object in the experiment exceeds binocular parallax. Further studies are needed, however, to determine how the stimulus produced by the quantitative relationship between motion parallax and binocular parallax influences the cardboard effect. In addition to movement of the object, we must investigate the effects of camera movement towards the object.

Next, we considered the interaction between binocular parallax (degree of thickness) and motion parallax (turning of the object). The significance of these experimental results is 5% and not, therefore, statistically significant (p-value: 0.05). It turns out, however, that lower thickness correlates with larger differences, depending on whether the object is turning or not. Motion parallax tends to dominate when the effects of binocular parallax are small; any effects due to other factors seem larger at such times. This appears to suggest the existence of nonlinearity in the processing of diverse factors (including binocular parallax, motion parallax, contours and shading) with regard to shape perception.

2.4.6   Summary

We have defined the cardboard effect in terms of the ratio of the magnification of the image's apparent size to its depth (thickness). When using the parallel camera configuration, the thickness is understood in terms of the ratio of the viewing angle of the lens (lens focal length) to the viewing angles of the screen or display. It is also associated with camera separation, convergence distance, and viewing distance when using the toed-in camera configuration. We have studied the correlation between thickness and subjective evaluations and found close agreement.

Besides binocular parallax, we also investigated the cardboard effect with regard to shape perception, looking at the influence of background and motion parallax for a moving object or camera. It has become evident with regard to the cardboard effect that binocular parallax (thickness) is the dominant factor, but the presence of background and motion parallax are also major factors. It is also clear that when the binocular parallax (thickness) is inadequate, motion parallax becomes a crucial factor for diminishing the cardboard effect.

2.5   Analysis of Distortion Free Conditions (Still Pictures)

2.5.1   Introduction

This section presents the subjective investigation of the orthostereoscopic condition. These are the basic conditions for understanding various features of 3-D images. As we saw in Fig. 2.1, the orthostereoscopic condition involves both recording and viewing conditions and can hardly be expected to be fulfilled in all actual broadcasting. Under the condition, however, when real space is projected onto the 3-D image space at a single magnification (as calculated theoretically); the distortions characteristic of 3-D images, including the cardboard and puppet theater effects, do not emerge as far as we can tell. We analyze the reproduced 3-D space subjectively in various recording conditions, looking at objects such as the mannequin standing in the hallway shown in Fig. 2.18, in order to demonstrate the absence of distortion. In Sect. 2.6, we analyze the reproduced 3-D image space in terms of "naturalness" and "unnaturalness" and examine various relevant factors.

Fig. 2.18 Mannequin used in the test

2.5.2 *Recording and Viewing Conditions*

For both the parallel and toed-in methods, the focal length of the lenses was set close to the average viewing angle of the lens and at 12 mm (43.6° viewing angle, which is the length most often used in actual production), and the cameras were placed 65 mm apart, which is the average distance between the human pupils. In the case of the parallel method, a camera separation of 65 mm is a necessary condition for Orthostereoscopy, as shown in Fig. 2.1. The object was stationary in the experiment, so we moved one camera horizontally to simulate a distance of 65 mm. We do, however, need two cameras, left and right, to shoot ordinary moving objects, so this distance of 65 mm cannot be obtained unless the cameras and their lenses are adapted for this purpose. We also, therefore, investigated distances of 75 mm and 85 mm with optical axes in parallel. The shooting distances to the mannequin were set at 1 m intervals from 2.5 to 6.5 m in five steps. These distances were obtained by moving the camera in order to avoid any subjective impressions of distance produced by moving the mannequin. The distance to the intersection of the optical axes was fixed at 4.5 m for the conventional recording method, which is equivalent to the viewing distance for a 160-inch screen.

We used three screen sizes: 160, 112, and 80 inch. The viewing distances for these screens were set at 4.5, 3.1, and 2.2 m, respectively, to make their viewing angles equal to the viewing angle of the lens (43.6°). The total shift for the left and right images (H_c in Fig. 2.1) using the parallel camera configuration was 65 mm. These conditions are outlined in Table 2.5.

Table 2.5 Recording and viewing conditions

Shooting target	Mannequin standing in the hallway
Shooting distance	2.5, 3.5, 4.5, 5.5, 6.5 m
Focal length of lens	12 mm
(Angle of view of lens)	43.6°
Camera separation:	
Toed-in	65 mm
Parallel	65, 75, 85 mm
Convergence point (toed-in)	4.5 m
Display size (viewing distance)	160 inch (4.5 m)
	112 inch (3.1 m)
	80 inch (2.2 m)
Total shift amount of L and R images	65 mm (in case of parallel method)
Display system	3-D-HDTV polarizing
Brightness	50.9 cd/m

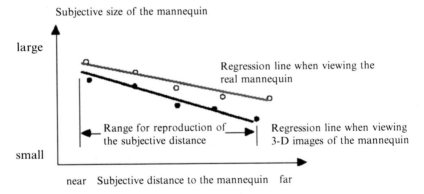

Fig. 2.19 Subjective reproduction of 3-D image space

2.5.3   Definition of Subjective Orthostereoscopy

The subjective distances to and sizes of the 3-D images of the mannequin shot from different distances were taken as the concrete criteria for subjective assessment to see whether the reproduced 3-D image space was orthostereoscopic. The outcomes were used to analyze the covariance with linear regression (covariance analysis) as shown in Fig. 2.19. Here, the 3-D image space should be considered subjectively orthostereoscopic if, in Fig. 2.19, the regression line, when compared with viewing of the real mannequin:

1. has the same gradient (parallelism of the regression plane), and
2. has the same height (homogeneity of regression).

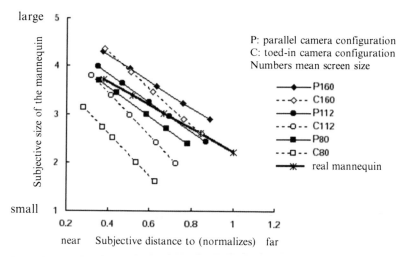

Fig. 2.20 Reproduction of the 3-D image space by parallel and toed-in camera configurations and screen size

2.5.4   Evaluation

To evaluate the test images obtained under these recording and viewing conditions and also the outcomes when viewing the real mannequin, we used the five-grade scale shown below for subjective assessment of the mannequin's size, and we measured the subjective distance of the mannequin distance in meters. The real mannequin was viewed from the camera distance. The subjective distance was normalized for each subject by the longest actual viewing distance. The scale used was: (1) small, (2) rather small, (3) normal, (4) rather large, and (5) large.

Each image was displayed for 15 s at 5-s intervals in darkness (the viewing distance was visible to some extent due to the light emitted from the projector). Test images were displayed at random with the same pictures each appearing twice. The subjects, 11 in all, were men and women in their 20s to 50s. All had been confirmed to have no problem in perceiving 3-D images.

2.5.5   Test Results

Just like in Fig. 2.19, Fig. 2.20 shows, with regression lines, the results for the parallel and toed-in camera configurations and viewing the real mannequin for all screen sizes. In Fig. 2.20, broken line C shows the results for the toed-in camera configuration and solid line P those for the parallel camera configuration with 65-mm separation. The geometric object configuration during recording and reproduction using this latter method is theoretically orthostereoscopic for all screen sizes but the results do

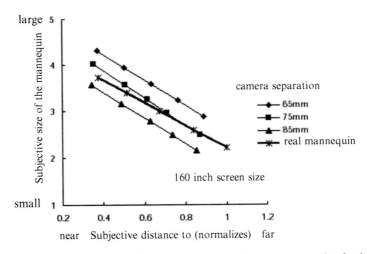

Fig. 2.21 Subjective reproduction of the 3-D image space with camera separation for the parallel camera configuration

Table 2.6 Results of parallelism test

Camera configuration	Screen (inch)	r^2	p-Value	Assessment	Common gradient
Parallel (65 mm)	160	0.99	0.26		−2.78
Parallel (75 mm)	160	0.98	0.12		−2.78
Parallel (85 mm)	160	0.99	0.22		−2.78
Toed-in (65 mm)	160	0.95	0.03	*	−4.12
Parallel (65 mm)	112	0.99	0.06		−2.78
Toed-in (65 mm)	112	0.95	0.01	**	−4.12
Parallel (65 mm)	80	0.97	0.11		−2.78
Toed-in (65 mm)	80	0.95	0.01	**	−4.12

p-Value probability
*5% significance; **1% significance

reveal some influence from screen size even so. Figure 2.21 shows the results using the parallel camera configuration and a 160-inch screen at camera distances of 65, 75, and 85 mm. The coefficients of determination (r^2) of all regression lines are more than 0.95.

Applying the definition of subjective Orthostereoscopy provided in Sect. 2.5.3, we hypothesize that the respective regression lines of reproduced images in the tests should be parallel to those for viewing of the real mannequin. Table 2.6 shows that this hypothesis cannot be rejected in the case of the regression lines for test images shot by the parallel camera configuration, including both the 75 and 85 mm camera separation modes, which do have the same gradient as when viewing the actual mannequin. The p-values are higher than 0.05 and there is a significance of 5%. The gradients do differ when the toed-in camera configuration is used. The hypothesis can be rejected if the p-value falls below 0.05 and the significance is 5%. It is clear from Table 2.7 that the regression lines do have a common gradient for viewing of the real

Table 2.7 Parallelism test by parallel optical axes

Variation	Square sum	Degree of freedom	Mean square	Fc	p-Value
Non-parallelism	0.053	5	0.011	1.48	0.25
Residual variation	0.130	18	0.007		

Table 2.8 Parallelism test by crossed optical axes

Variation	Square sum	Degree of freedom	Mean square	Fc	p-Value
Non-parallelism	0.031	2	0.016	0.45	0.65
Residual variation	0.314	9	0.035		

Table 2.9 Test for homogeneity of regression by parallel optical axes

Camera configuration	Screen (inch)	Y-intercept	T-value	Assessment
Parallel (65 mm)	160	5.35	8.36	**
Parallel (75 mm)	160	4.95	1.40	
Parallel (85 mm)	160	4.52	−6.21	**
Parallel (65 mm)	112	4.91	0.61	
Parallel (65 mm)	80	4.62	−4.41	**

**1% significance

mannequin and shooting with the parallel camera configuration (p-value: probability of 0.25), and from Table 2.8 that images shot by the toed-in camera configuration can also be represented by a common gradient (p-value: probability of 0.65).

Table 2.9 shows results of the test for the homogeneity of regression of test images shot by the parallel camera configuration. This table suggests that the heights of the regression lines diverged from those for actual viewing of the mannequin with 65 mm camera separation using a 160-inch screen, 65 mm camera separation with an 80-inch screen and 85 mm camera separation with a 160-inch screen. For the toed-in camera configuration, conversely, there is no sense in testing for the homogeneity of regression because the regression line gradients are different for actual viewing and test images. The comparison of images shot by this camera configuration does, however, show that the heights of regression lines differ distinctly with screen size, as shown in Fig. 2.20 (p-value $= 9.4E-28$).

2.5.6 Discussion

2.5.6.1 Viewing a Real Mannequin

In terms of the constancy of size, the regression line when the actual mannequin is viewed should have no gradient. Constancy of size is not, however, obtained in all cases and this may be taken to have been the case in this experiment, also.

2.5.6.2   Parallel Camera Configuration (the Orthostereoscopic Condition)

Theoretically, cameras placed 65-mm apart in the parallel camera configuration produce orthostereoscopic effects whereby 3-D space should be reproduced in just the same way as real space on screens of any size. In fact, the regression lines plotted out for this camera configuration in Fig. 2.20 have the same gradient as the regression lines when viewing the real mannequin but different heights depending on screen size (see Table 2.9). In the cases of the 160- and 80-inch screens, it cannot be said that the 3-D images are subjectively orthostereoscopic. At the same time, the range of reproduction of subjective distance (subjective distances of the mannequin perceived to be farthest and nearest) is significantly small. Put another way, the viewer can perceive subjective space that has been evenly reduced from its actual size or distance with smaller screen sizes (in this case, the viewing distance is also shortened in order to keep the viewing angle for the screen constant at 43.6°) even under the orthostereoscopic condition. This reduction effect, however, is due to the difference in regression line heights and may be different from the reduction effect in the case of the toed-in camera configuration. We will return to this issue in Sect. 2.5.6.4. The regression lines of test images reproduced on a 112-inch screen (2.5 m wide) are identical both in gradient and height to the regression lines of real mannequin viewing. The orthostereoscopic space is, accordingly, reproduced subjectively. These results may be taken to corroborate the argument that there is a minimum screen size for preventing the puppet theater effect [14] and that a screen wider than 2 m is needed to reproduce images unaffected by the reduction effect [8]. It is to be noted, however, that the range of reproduction of the subjective distance is about 90% of that for real viewing. We may perhaps have to give further consideration to the reproduction of real space sensations of distance in 3-D images.

2.5.6.3   Toed-In Camera Configuration

As we saw in Fig. 2.20, there is a common gradient with the toed-in camera configuration (see Table 2.8). The large differences of screen size and differences in the heights of regression lines are reflected in the range of reproduced subjective distance. The toed-in camera configuration always reproduces an object at the optical axis intersection of the screen position regardless of screen size or viewing distance and forms the 3-D image space before and behind the object. In the experiment described here, where the screen size is changed and the viewing angle is fixed in relation to the screen, the difference of screen size is mirrored directly in the subjective size and distances of the 3-D image space to be reproduced. The toed-in experiment shown in Fig. 2.20 may be taken to indicate that the difference in the physical size of the viewing space, which varies analogously, is reflected in the subjective expansion of the 3-D image space.

2.5.6.4 Gradient Differences of Parallel/Toed-In Regression Lines

As shown in Fig. 2.20, the regression lines of images shot by the toed-in camera configuration had different gradients from those of the regression lines obtained by the parallel camera configuration and actual viewing, regardless of screen size. This renders any test for the homogeneity of regression meaningless, so we will only discuss the gradients themselves here. As a typical example of the toed-in method, we attach importance to the test results of the 160-inch screen where the viewing distance is equivalent to the distance to the optical axis intersection of the cameras. Mathematically, the size of the mannequin on the screen (or on the retinas) and the position where 3-D images are produced should be the same as those of the parallel camera configuration complying with the orthostereoscopic condition. In the experiment, however, use of the toed-in method caused the gradients to differ from those obtained by the parallel method, as we saw in Fig. 2.20. This figure also indicates that the difference in subjective size for long subjective distances has a major influence on the difference of gradient. In other words, the mannequin's image produced by the toed-in camera configuration is perceived to be smaller for longer shooting distances in comparison to images produced by the parallel camera configuration. This can be explained as follows: even though the camera separation is 65 mm, which is the average distance between the observer's left and right irises, the puppet theater effect will arise with ordinary images when the background is shot by the toed-in camera configuration as opposed to the parallel camera configuration (in compliance with the orthostereoscopic condition), due to discrepancies between the background depth information of lens perspective and binocular parallax. The puppet theater effect is intensified by these discrepancies in proportion to the greater shooting distance and the percentage of background area shown on the screen. This is why tests using the parallel camera configuration, which generates no such discrepancy, do not produce the difference in regression line gradients mentioned in Sect. 2.5.6.2.

2.5.6.5 Changing Parallel Camera Separation

As seen in Fig. 2.21, no difference was observed between the regression line gradients of 3-D images shot by cameras separated by 65, 75, or 85 mm in the parallel camera configuration and those of actual mannequin viewing. There were large differences in the heights of regression lines, with the subjective sizes of the objects decreasing with wider camera separation, suggesting that the distance between cameras is a very important parameter for 3-D image production. There is no significant difference in the range of subjective distance reproduction for camera separations of 65 and 85 mm.

2.5.7 Summary

In the case of 3-D images shot and displayed by the parallel camera configuration, the heights of regression lines of subjective size with regard to the subjective distance

Table 2.10 Recording and viewing conditions

Program content	Softball game
Camera position	4.5 m diagonally behind home plate and to its right
Focal length of lens	12 mm
(Viewing angle of lens)	43.6°
Camera separation toed-in parallel	65 mm
Convergence point (toed-in)	4.5 m
Display size (viewing distance)	160 inch (4.5 m)
	112 inch (3.1 m)
	80 inch (2.2 m)
Total shift amount of left and right images	65 mm (in case of parallel method)
Display system	3-D HDTV polarizing
Brightness	50.9 cd/m

may differ from those seen with actual mannequin viewing, and this is due to screen size. The regression line gradients of 3-D images are apparently the same as actual mannequin viewing gradients. The regression line gradients of 3-D images shot by a toed-in camera configuration differ from those of 3-D images shot by a parallel camera configuration, or when a real mannequin is viewed. There are distinct differences by screen size and regression line heights; and this difference of gradient produces the puppet theater effect in images with background.

2.6 Analysis of Distortion Free Conditions (Moving Images)

2.6.1 Introduction

In Sect. 2.5, we generated regression lines for subjective object size with regard to the subjective distance to a stationary object, and analyzed subjective reproduction of 3-D image space by comparing the gradients and heights of regression lines when the actual mannequin was viewed with the gradients and heights of the regression lines for the viewing of 3-D images. Here, we examine how these gradient and height differences affect the naturalness of the stereoscopic effects of 3-D HDTV programs produced using the parallel and toed-in camera configurations.

2.6.2 Recording and Viewing Conditions

As shown in Table 2.10, the recording and viewing conditions are the same as in the experiment outlined in Table 2.5. The focal length of the lens used is 12 mm (at a lens viewing angle of 43.6°); camera separation for both parallel and toed-in camera configurations is 65 mm; distance to the convergence point in the case of the toed-in camera configuration is 4.5 m; and display screen sizes are 160, 112, and 80 inch.

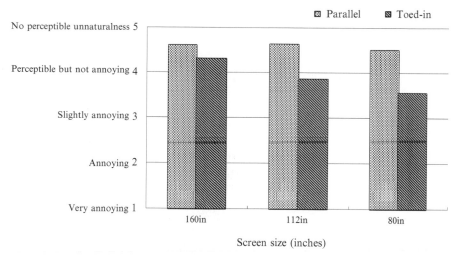

Fig. 2.22 Evaluation of the naturalness of 3-D images

The program shows a softball game. The camera position is 4.5 m diagonally behind home plate and to its right.

2.6.3 Evaluation Test

Under these conditions, we took four cuts from the program for use as test images: two from the parallel version and two from the toed-in version. The subjects were requested to assess the reproducibility of the 3-D image space (visual depth) in terms of object size and 3-D sensation using the following scale: (1) very annoying; (2) annoying; (3) slightly annoying; (4) perceptible, but not annoying; and, (5) no perceptible unnaturalness.

Each cut was shown for 25 s with 5-s intervals in between. They were presented at random and each cut was shown twice. The attributes of the subjects were the same as in the first experiment, but the number was reduced to ten. All of the subjects had real experience of playing or watching softball and were familiar with the situations.

2.6.4 Experimental Results

Figure 2.22 tabulates the results of the experiment. The variance analysis in Table 2.11 shows significant differences between the parallel and toed-in camera configurations and between display screen sizes. The scores for all 3-D images shot

Table 2.11 Variance analysis

Factor	Square sum	Degree of freedom	Mean square	F-value	p-Value
Screen size	3.61	2	1.81	3.58	0.03*
Camera configuration	16.67	1	13.67	27.13	0.00**
Interaction	2.26	2	1.13	2.25	0.11
Error	57.44	114	0.50		
Total	76.98	119			

*5% significance; **1% significance

Table 2.12 Parameters for multiple regression analysis

Camera configuration	Screen size (inch)	Gradient	Line of gravity	Reproduction range	Evaluation	Estimate
Parallel	160	−2.95	3.25	0.52	4.60	4.67
Parallel	112	−3.00	3.25	0.52	4.63	4.65
Parallel	80	−3.06	3.04	0.43	4.50	4.44
Toed-in	160	−3.77	3.43	0.47	4.30	4.29
Toed-in	112	−4.40	2.90	0.41	3.85	3.81
Toed-in	80	−4.39	2.39	0.35	3.55	3.59
Actually used		−2.43	2.98	0.62	5.00	4.97

by the parallel camera configuration were within the limit for imperceptibility (score of 4.5), meaning that the images looked natural to the subjects. All 3-D images produced by the toed-in camera configuration were considered acceptable, but the scores tended to be worse with smaller screen sizes.

2.6.5   Discussion

In order to assess which factors contributed to this unnaturalness, we performed multiple regression analysis of the test results with unnaturalness as an objective variable and the differences in regression line gradients and heights, plus the reproduction range for subjective distances (Fig. 2.20) as explanatory variables. Here, the height of the regression line was the line's center of gravity. Table 2.12 shows the explanatory variables of the regression lines of the test images. The determination coefficient to express the quality of fit of the multiple regression equation was 0.99. The p-value (probability) for regression was 0.0014, and it was significant. It is clear, therefore, that this multiple regression equation has a high degree of analytical precision and should be an effective way to gauge the factors involved in unnaturalness. Table 2.13 shows the standard partial regression coefficients of the explanatory variables obtained using this equation. The table shows the partial regression coefficient of the regression line gradient to be prominent at 0.67, which suggests that this is a major cause of the unnaturalness of 3-D images. In other words, in order to produce natural 3-D images, it is important to reproduce the rate of change of the subjective object size to subjective object distance that is obtained in actual

Table 2.13 Standard partial regression coefficients of the multiple regression equation

Variable name	Partial regression coefficient	Standard partial regression coefficient
Gradient	0.44	0.67
Center of gravity	0.26	0.18
Reproduction range	1.35	0.24
Constant	4.40	

viewing. It follows that it is important to reproduce the perspective of the lens (linear perspective) naturally (using a standard lens or the like) and, at the same time, maintain the proper balance between the sense of depth obtained from perspective and that obtained by means of binocular parallax. The orthostereoscopic condition is considered to apply in this shooting method.

2.6.6   Summary

We have here discussed the subjective reproduction of 3-D image space obtained when shooting real objects and their backgrounds. From the viewpoint of object size and stereoscopic effects, we further studied the sensations of naturalness and unnaturalness, paying attention not only to the differences between the traditional shooting method and a shooting method that satisfies the orthostereoscopic condition but also the differences produced on different screen sizes. We have seen that although the subjective size of an object and the reproduction range of the subjective distance are influenced by screen size under orthostereoscopic conditions, the rate of change of subjective object size to subjective object distance is the same as in real viewing for all screen sizes. It was also found that this is an essential factor for 3-D images to seem natural to the viewer. In the case of the traditional shooting method, however, it became apparent that the rate of change of subjective object size to subjective object distance is larger than during actual viewing, and this increases the likelihood that the puppet theater effect will occur. It is also clear that the difference in screen sizes has a significant impact on the subjective size of an object and the reproduction range of the subjective distance.

The creation of natural 3-D images for broadcast requires not only the use of standard lenses that neither stress the perspective nor reduce it, and lenses with almost the same focal length as each other, but also maintenance of a balance between depth perception by perspective and depth perception by binocular parallax. The orthostereoscopic condition we have adopted in this analysis is considered to provide an effective shooting method. In terms of actual program production and the viewing environment, however, we are aware that the orthostereoscopic condition is not always fulfilled. As a practical matter, it is necessary to take advantage of the characteristics of parallel and toed-in camera configurations, including the orthostereoscopic condition, to produce programs.

References

1. A. J. Hill, "A Mathematical and Experimental Foundation for Stereoscopic Photography", Journal of SMPTE, Vol. 61, pp. 461-486 (1953)
2. E. Levonian, "Stereography and Transmission of Images", Journal of SMPTE, Vol. 64, pp. 77-85 (1955)
3. A. Woods, T. Docherty, R. Koch, "Image Distortions in Stereoscopic Video Systems", Proc. of SPIE Vol.1915, Stereoscopic Displays and Applications IV (1993)
4. A. M. Ariyaeeinia, "Analysis and design of stereoscopic television systems", Signal processing: Image Communication 13, pp. 210-208 (1998)
5. H. Yamanoue, M. Okui, F. Okano, "Geometrical Analysis of Puppet Theater and Cardboard effects in Stereoscopic HDTV Images", IEEE Transactions on Circuits and Systems for Video Technology, Vol.16, No.6 (June. 2006)
6. R. Spottiswoode, N. L. Spottiswoode, C. Smith, "Basic Principles of the Three Dimensional Film",SMPTE J. 59,pp. 249-286 (Oct. 1952)
7. H. Yamanoue, M. Nagayama, M. Bitou, J. Tanada, "Orthostereoscopic conditions for 3-D HDTV" Proc. of SPIE Vol. 3295, Stereoscopic Displays and Virtual Reality Systems V (1998)
8. C. W. Smith, A. A. Dumbreck, "3-D TV: The Practical Requirements", Television J. of Royal Television Society, pp. 9-15 (Jan./Feb. 1988)
9. S. Pastoor, "Human factors of 3-D imaging: Results of recent research at Heinrich-Hertz-Institute Berlin", Proc. ASIA Display'95 Conf. (1995)
10. T. Komatsu, S. Pastoor, "Puppet theater effect observing stereoscopic images", Technical Report of IEICE, IE 92-104, pp. 39-46 (1993)
11. H. Yamanoue, "The relation between size distortion and shooting conditions for stereoscopic images", SMPTE J. pp. 225-232, April 1997
12. K. Hopf, "An Autostereoscopic Display Providing Comfortable Viewing Conditions and a High Degree of Telepresece", IEEE Transactions on Circuits and Systems for Video Technology, Vol.10, No.3 (Apr. 2000)
13. D.L. MacADAM,: "Stereoscopic Perceptions of Size, Shape, Distance and Direction", SMPTE J. 62, pp. 271-289 (1954)
14. S. Herman, "Principles of Binocular 3-D Displays with Applications to Television", SMPTE J. 80, pp. 539-544 (July 1971)
15. Lydia M. J. Meesters, Wijnand A. Ijsselsteijn, Pieter J. H. Senuntiens, "A Survey of Perceptual Evaluations and Requirements of Three-Dimensional TV", IEEE Transactions on Circuits and Systems for Video Technology, Vol. 14, No. 3 (Mar. 2004)

Chapter 3
Research on Differences Between the Characteristics of Left and Right Images

Abstract Many stereoscopic imaging systems, including the 3-D HDTV system, use separate devices for the left and right eyes. It is difficult, however, to integrate the characteristics of the left and right images fully due to factors deriving from device performance, and inconsistencies remain. In this chapter, we inquire into the acceptable margins for differences of size, vertical shift, rotation, brightness, and Crosstalk between left and right images and draw up a parameter index for the display of easy-to-watch 3-D images.

Keywords 3-D HDTV • Black clip • Crosstalk • Geometrical distortion • Guidelines for 3-D camera adjustment • Rotation error • Size inconsistency • Stereoscopic HDTV • Vertical shift • White clip

3.1 Inconsistencies of Size, Vertical Shift, and Rotation

3.1.1 Introduction

Geometrical inconsistencies between left and right images arise when the left and right cameras are placed asymmetrically. A vertical shift arises if one of the cameras tilts up or down, a rotational error if one of the cameras tilts horizontally, and a difference of size if the focal lengths of the left and right lenses are not the same. There have been studies on how camera calibration errors and vertical shift affected depth perception [1], the scope of fusion using random-dot patterns [2], and acceptable margins for film and traditional television systems [3, 4], but not on the acceptable margins for HDTV. Although complex geometrical distortion may occur at the time of shooting, we have not in fact encountered any examples of this. What happens in practice is that a lot of time is spent adjusting the focal length of

Table 3.1 Added geometrical distortions

Step	Rotational error (°)	Size gap (%)	Vertical shift (%)
1	0.00	0.00	0.00
2	0.72	1.50	0.89
3	1.44	3.00	1.78
4	2.16	4.50	2.67

the lenses and positions of the left and right cameras to eliminate the geometrical distortions in program production. At times, the adjustable features become so bulky that the 3-D cameras themselves are huge, thereby losing the advantage of compactness. This chapter considers the tolerance for geometrical distortion, adjustment of the optical axes, and optimization of the adjustment mechanisms for the efficient production of 3-D HDTV programs.

3.1.2   Geometrical Distortions in Experimental Situations

These experiments added geometrical distortion artificially in order to grasp distortions in a quantitative manner. The distortions included rotational errors from horizontal inconsistencies, size inconsistencies due to differences of focal length, and vertical shifts due to camera tilt. We created these distortions using HDTV digital video effect (DVE).

Table 3.1 shows the amounts of geometrical distortion that were added. Rotational error refers to relative rotation of the two images, namely, the total rotation to the left of the L image and to the right of the R image in degrees. The size gap refers to the effects of zooming in and out; this gap represents the total zooms of the L and R images expressed as a percentage ratio. The vertical shift represents the total upward shift of the L image against the downward shift of the R image, expressed as a percentage ratio of screen height.

3.1.3   Experiment 1 (Differences Between Pictures)

The acceptable margins of distortion may vary from picture to picture. Here, we examined not only acceptable margins for distortion, but also the differences picture by picture. We used four 3-D images, three general images containing landscapes or objects in the background, and one CG image with a brick/tile pattern. These pictures were chosen from among recent 3-D HDTV broadcasts. The general images contained near-to-distant views for various coverage situations (i.e. the images had items positioned in the foreground and background).

Image A: a coconut palm tree

Image C: Miyako Harumi

Image B: a woman on the horse

Image D: red brick

Fig. 3.1 Evaluating the images used in Experiment 1

Image A shows a near/intermediate view (from a coconut palm at 0.3-D to 0.1 D, where D is the inverse of the distance, expressed as 1/m), and a distant view of background. Image B consists of a near view (trees in the foreground at 0.2 D), an intermediate view (a woman on the horse at 0.1 D), and a distant view (background). Image C consists of a near view (stage at about 0.25 D), an intermediate view (a singer at about 0.15 D), and a distant view (stage set). These images are shown in Fig. 3.1.

Showing the subjects the original pictures for 10 s and then displaying the test images for 10 s, we collected evaluations on a scale of 1–5, as follows: (1) very annoying; (2) annoying; (3) a problem, but not annoying; (4) distortion perceived, but not a problem; and (5) distortion not perceived.

Prior to the experiment, we explained to the subjects that "a problem" meant "feeling bothered by distortions of the reproduced image space" due to geometrical distortion, and that "annoying" meant "feeling annoyance with the state of binocular fusion" beyond merely "feeling bothered by the distortions of the reproduced image space." The ten subjects were men and women aged from their 20s to 50s, they were all very familiar with 3-D images. Each subject completed the experiment twice. The screen was 160 inch (355 cm wide × 200 cm high) and the viewing distance about 6 m (3 H, where H is screen height). We separated the left and right images by polarization filter and polarized glasses.

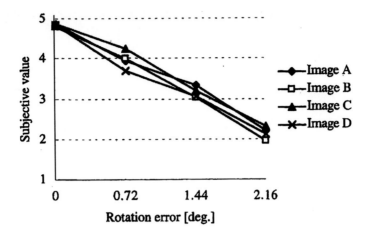

Fig. 3.2 Rotational error (degrees)

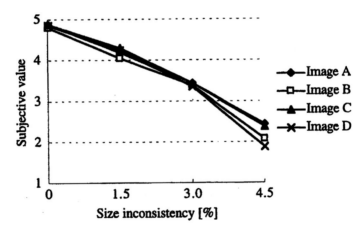

Fig. 3.3 Size inconsistency (%)

3.1.4   Results of Experiment 1

Figures 3.2, 3.3, and 3.4 show the results for rotational error, size inconsistency, and vertical shift, respectively. In each figure, the vertical axis represents the added distortions and the horizontal axis the subjective evaluations, with the differences between pictures as parameters. Variance analysis of the rotational errors, size inconsistencies and vertical shifts are shown in Tables 3.2, 3.3, and 3.4, respectively. The results show no difference regarding picture type and geometrical distortion by variance analysis. We next examined the differences in mean value (least significant difference method) to see if any difference between the pictures emerged in the subjective evaluations. We did not find any significant difference (risk rate: 1%) between

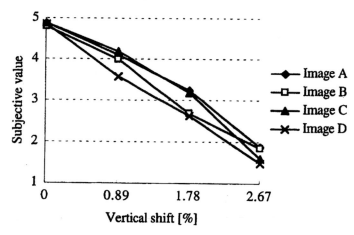

Fig. 3.4 Vertical shift (%)

Table 3.2 Analysis of variance (rotational error)

Factor	Sum of squared deviations	Degree of freedom	Mean square	f-number
A: picture	3.02	3	1.01	2.27
B: inclination	319.27	3	106.42	240.36**
A×B	2.59	9	0.29	0.65
Error	134.60	304	0.44	
Total	459.47	319		

**1% significant

Table 3.3 Analysis of variance (size inconsistency)

Factor	Sum of squared deviations	Degree of freedom	Mean square	f-number
A: picture	2.29	3	0.76	1.54
B: size	312.69	3	104.23	210.85**
A×B	2.79	9	0.31	0.63
Error	150.28	304	0.49	
Total	468.05	319		

**1% significant

Table 3.4 Analysis of variance (vertical shift)

Factor	Sum of squared deviations	Degree of freedom	Mean square	f-number
A: picture	7.47	3	2.49	4.66**
B: verticality	441.38	3	147.13	275.28**
A×B	5.68	9	0.63	1.18
Error	162.48	304	0.53	
Total	617.00	319		

**1% significant

Table 3.5 Detection threshold and acceptable margin of geometrical distortion

	Rotational error (°)	Size inconsistency (%)	Vertical shift (%)
Detection threshold	0.53 ± 0.22	1.21 ± 0.30	0.66 ± 0.49
Acceptable margin	1.14 ± 0.16	2.89 ± 0.21	1.45 ± 0.28

images with regard to size inconsistency, but did find a significant difference (risk rate: 1%) between the general images (Images A and C) and the CG image (Image D) with regard to vertical shift. It is presumed that a vertical shift is detected more easily in the image with a brick/tile-like lattice pattern than in the general images. As the "f-number" in Table 3.4 suggests, the effect due to differences between the pictures is smaller than that caused by the added geometrical distortion. We did find significant differences (risk rate: 1%) between all of the steps shown in Table 3.1 with regard to rotational errors, size inconsistency, and vertical shifts. Based on the above, we show the acceptable margins of geometrical distortion in Table 3.5.

3.1.5   Experiment 2 (Distortions Combined)

In actual shooting, the above-mentioned geometrical distortions may occur in combination. We performed variance and multiple regression analyses to determine how the combination of geometrical distortions affects subjective evaluations.

We used Images B and C from among the general images used in Experiment 1, adding a total of 27 kinds of distortion ($3 \times 3 \times 3$) by combining the distortions shown in Table 3.6 in various ways. It is to be noted that while the distortions were added in the same way as in Experiment 1, the combinations were made in the order of rotational error, size inconsistency, and vertical shift. As in Experiment 1, the evaluations were made according to a five-point scale.

The subjects were men and women aged from their 20s to 50s; all subjects were very familiar with 3-D images. For Image B, we ran the experiment with 11 subjects in triplicate. For Image C, we ran the experiment with 13 subjects in duplicate. Other conditions were the same as in the previous experiment.

3.1.6   Results of Experiment 2

3.1.6.1   Variance Analysis

Figures 3.5 and 3.6 show the results for Images B and C, respectively. Both show vertical shift on the horizontal axis and take rotational error as the parameter. The variance analyses for rotational error and vertical shift are shown in Tables 3.7 and 3.8, respectively. We found that each distortion remained a significant factor for both images, even in combination. In addition, interaction between the added distortions

Table 3.6 Added geometrical distortions

Step	Rotational error (°)	Size gap (%)	Vertical shift (%)
1	0.00	0.00	0.00
2	0.72	1.50	0.89
3	1.44	3.00	1.78

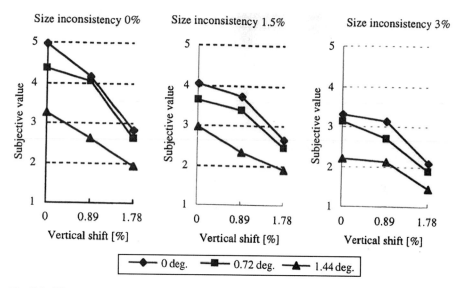

Fig. 3.5 When geometrical distortions are combined (Image B)

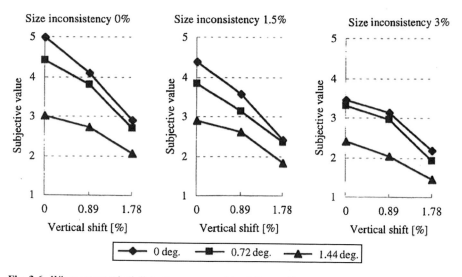

Fig. 3.6 When geometrical distortions are combined (Image C)

Table 3.7 Variance analysis (Image B)

Factor	Sum of squared deviations	Degree of freedom	Mean square	f-number
A: inclination	200.81	2	100.41	174.28**
B: verticality	281.32	2	140.66	244.16**
C: size	139.44	2	69.72	121.02**
A×B	10.48	4	2.62	4.55**
A×C	6.43	4	1.61	2.79*
B×C	12.28	4	3.07	5.33*
A×B×C	4.91	8	0.61	1.06
Error	497.76	864	0.58	
Total	1,153.42	890		

*5% significant; **1% significant

Table 3.8 Variance analysis (Image C)

Factor	Sum of squared deviations	Degree of freedom	Mean square	f-number
A: inclination	157.58	2	78.79	143.26**
B: verticality	246.23	2	123.11	223.85**
C: size	89.88	2	44.94	81.71**
A×B	12.23	4	3.06	5.56**
A×C	5.40	4	1.35	2.45*
B×C	3.42	4	0.86	1.56
A×B×C	3.76	8	0.47	0.85
Error	371.24	675	0.55	
Total	889.73	701		

*5% significant; **1% significant

was noted for both images. This pattern seems to be related to the fact that some geometrical distortions did not stand out on their own, depending on how they were generated or added, and features of the pictures themselves had an effect. As suggested by the f-numbers, the effect due to interactions was smaller than the main effect.

3.1.6.2 Multiple Regression Analysis No. 1

The variance analysis revealed interactions between the added geometrical distortions when geometrical distortions were combined and that the main effects of rotational error, size inconsistency, and vertical shift were major factors. We therefore used these as explanatory variables in the multiple regression analysis and subjective evaluation as an objective variable.

We obtained these regression equations for Images B and C ((3.1) for B and (3.2) for C, respectively):

$$S = -0.78(R) - 0.32(S) - 0.75(V) + 4.68 \qquad (3.1)$$

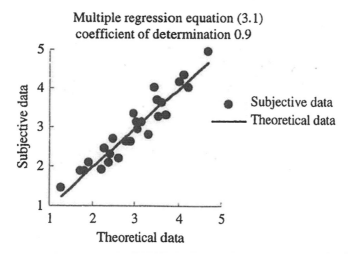

Fig. 3.7 The estimated values computed with the regression equation (3.1) and observed values for Image B

$$S = -0.77(R) - 0.29(S) - 0.80(V) + 4.71 \qquad (3.2)$$

In these equations, S stands for the estimated subjective evaluation, (R) for rotational error (degrees), (S) for size inconsistency (%), and (V) for vertical shift (%). Figure 3.7 shows the estimated values computed with the regression equation and observed values for Image B.

The coefficient of determination that indicates the precision of analysis in the regression equation is 0.90 for Image B and 0.92 for Image C. Tests obtained p-values of 5.98E−12 and 9.71E−13 for Images B and C, respectively, which are both significant. It is clear that the above regression equations are effective formulas for predicting the detection of geometrical distortions in the experiments using each image. Comparing the regression coefficients, we can see that the influence produced by rotational error and vertical shift is almost the same, and the influence of size inconsistency is smaller.

3.1.6.3   Multiple Regression Analysis No. 2

In actual shooting, rotational error, size inconsistency, and vertical shift do not necessarily occur in the same order as in this experiment. Accordingly, even if the amount of geometrical distortion is fixed for rotational error, size inconsistency, and vertical shift, the interactions between those distortions (shown in Tables 3.7 and 3.8) have to be considered in terms of how and in what order those

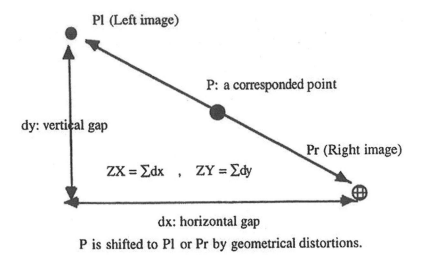

P is shifted to Pl or Pr by geometrical distortions.

Fig. 3.8 Vertical and horizontal shift due to geometrical distortion

distortions are created when they are fairly large. In the case of rotational errors, the center is not always the center of the image, as it is dependent on physical structures (such as the 3-D cameras and camera platforms). In order to produce a more broadly useful model, we performed multiple regression analysis to detect geometrical distortions according to the horizontal and vertical shifts deriving from added geometrical distortions on corresponding points on the screen. Figure 3.8 shows an example of shift occurring due to geometrical distortions on the screen at corresponding points. This figure shows how corresponding point P moves to Pr and Pl due to the distortions.

As explanatory variables, we analyzed the sums of vertical shift ($ZY = \Sigma dy/1e$-7: in pixels) and horizontal shift ($ZX = \Sigma dx/1e$-7: in pixels) obtained by integrating the absolute values of dx and dy over the entire screen, as shown in Fig. 3.8. The multiple regression equations obtained for Images B and C are shown in (3) and (4), respectively. In both equations, S is the estimated subjective evaluation.

$$S = -0.32\left(ZX\right) - 0.65\left(ZY\right) + 5.18 \tag{3.3}$$

$$S = -0.27\left(ZX\right) - 0.69\left(ZY\right) + 5.22 \tag{3.4}$$

Figure 3.9 shows the estimated values computed with the regression equation and the observed values for Image B. Tests produced p-values of 6.46E−16 and 6.03E−18 for Images B and C, respectively, and the foregoing equations were significant. The coefficient of determination for Image B is 0.95 and that for Image C is 0.96. The two coefficients, if converted into correlation coefficients,

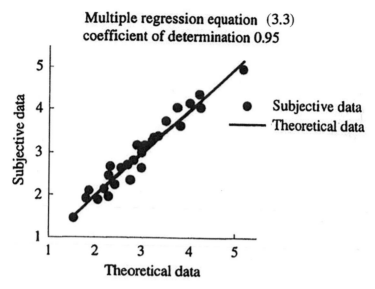

Fig. 3.9 The estimated values computed with the regression equation (3.3) and observed values for Image B

are 0.97 and 0.98. In addition, although the number of explanatory variables fell to two from three, these equations indicate more precise analysis than in the previous test.

The assumption of acceptable margins for horizontal and vertical shift due to added geometrical distortions at corresponding points on the screen produces more precise analysis than the assumption of acceptable margins taken directly from the amount of geometrical distortion actually added, and that the results obtained do reflect most of the observed data. In this case, geometrical distortions are not dependent on how they occur, the order of occurrence, the 3-D cameras or the camera platforms. Equations (3) and (4) are, therefore, regarded as useful practical models.

3.1.6.4   Testing Parallelism and Positional Gaps

We intended to generalize multiple regression equations (3) and (4) further with regard to the images shown above. We assumed that all constant terms of a regression equation are equal when all partial regression coefficients are the same (null hypothesis H01) and that null hypothesis H01 should not be rejected (null hypothesis H02). We tested parallelism and positional gaps to see if any common multiple regression equation exists regardless of which pictures are used. It turned out, for the test of parallelism, that null hypothesis H01 should not be rejected (at an f-number of 0.786 and p-value of 0.462). For positional gaps, null hypothesis H02 should not be rejected (at an f-number of 0.325 and p-value of 0.571). These multiple regression equations can, therefore, be regarded as a single common expression, which we state as Eq. (5).

$$S = -0.30(ZX) - 0.67(ZY) + 5.2 \qquad (3.5)$$

In this multiple regression equation, the p-value is 1.13E−34 (and significant) and the coefficient of determination is 0.95. Comparing the partial regression coefficients, we can see that the horizontal shift is more significant than the vertical effect. In our opinion, this corresponds to the finding by Isono et al. [2] that vertical shift has a narrower range of fusion than horizontal parallax.

3.1.7   Guidelines for Camera Adjustment

It is clear from Eq. (5) that adjustment to suppress ZY reduces geometrical distortions. We next analyzed the relationship between the assumed geometrical distortions and ZX and ZY based on real conditions in 3-D HDTV program production.

When we view images in the form of a television program, whether HDTV images or conventional television or film images, the size of the picture in every cut and relationship between picture sizes is very important. It follows from this that the camera position and lens selection are critical factors when shooting programs. In the case of 3-D HDTV, however, the selection of camera positions is limited by the size and mobility of the 3-D cameras, and cut-by-cut changes inevitably depend on use of the zoom lens in most cases. Zooming typically produces global inconsistencies between the left and right images because the optical axis of each zoom lens does not conform fully to the physical center of the image pickup device. Some horizontal shift is acceptable as parallax but the vertical shift causes problems as a geometrical distortion. The zoom differs for the left and right lenses and this difference emerges as a size inconsistency.

Conversely, rotational errors do not occur frequently if the level of the cameras is adjusted in the beginning, and certainly not as frequently as vertical shifts and size inconsistencies. We first focused, therefore, on the occurrence of vertical shifts. Figures 3.10 and 3.11 show the relationship between geometrical distortions and ZX and ZY. In the respective figures, the negative values for vertical shift and size inconsistency represent cases where the right image is smaller than the left image, or where downward shift occurs. We can see from the figures that when rotational errors are not factored into the calculation, ZX is nearly proportionate to the size inconsistencies. It is also evident that, with regard to ZY, vertical shifts are the dominant factor and size inconsistencies become influential when vertical shifts do not occur.

We cannot discuss rotational errors quantitatively because the center of rotation varies according to the geometry of each 3-D camera and structure of the camera

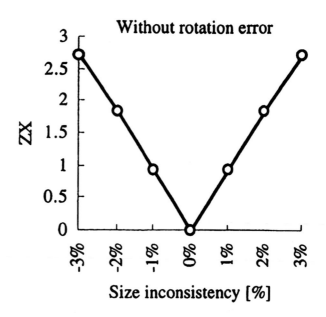

Fig. 3.10 Geometrical distortion and horizontal shift (ZX)

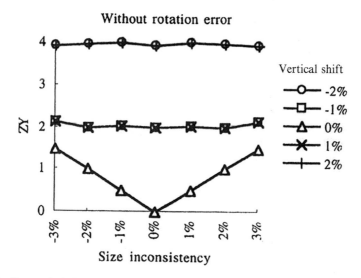

Fig. 3.11 Geometrical distortions and vertical shifts (ZY)

platform. Even so, it is clear that the mechanism of rotational error has a major impact on ZX and ZY. We suggest, therefore, that to reduce the geometrical distortions in 3-D HDTV shooting, it is most efficient first to secure the level of the cameras, adjust the vertical shifts that have such a large effect on ZY, and only then adjust the image size.

3.1.8   Summary

We have sought acceptable margins for handling geometrical distortions occurring
separately with regard to various images selected from among recent 3-D HDTV
programs, and found no significant differences for general images. In the light of
actual program shooting, we next studied cases in which geometrical distortions
were combined, focusing on general images. A model that assumes acceptable margins for vertical and horizontal shifts using corresponding points on the screen has
broader utility and higher analytical precision than one that assumes acceptable
margins directly from the geometrical distortions actually added. We also tested
parallelism and positional gaps in order to generalize the models obtained for each
image, and showed that these models could be integrated into a single one. Lastly,
we discussed guidelines for the efficient placement and adjustment of the left and
right cameras in program shooting on the basis of the models obtained.

3.2   White/Black Clip Levels

3.2.1   Introduction

The characteristic differences in brightness that arise between the left and right
images in the course of production are relatively easy to compensate for at the
receiver side with regard to luminance and contrast. It is more difficult to compensate for non-linear differences, however, including when the image is clipped at
certain white/black levels. In this section, therefore, viewing difficulty is evaluated
subjectively by experiments which examine how much disturbance is caused by
differences between the black/white clip levels of the left and right images.

3.2.2   Experimental Assessment Using Geometric Patterns

Three cases were tested in which either the left or right image has been clipped at the
camera and/or display. Case A applies when either of the images is clipped at the
display; Case B applies when the image is clipped during shooting, and then displayed
without adjustment of peak luminance levels of the left and right pictures; and Case-C
applies when the image is clipped during shooting, and then displayed after adjusting
the peak luminance levels for the left and right pictures. The amount of disturbance
was evaluated subjectively in each case by setting the white level clip as L_w and the
black clip level as L_b. A geometric pattern was used to exclude dependency on picture
designs, allocating quadrant blocks (viewing angle of about 60 arcmin) at various
random luminance levels so that the luminance-distribution histogram evened out
across the entire image. A time-division stereoscopic display device was also used.

Figure 3.12 shows the averaged results of evaluations performed by seven individuals. As observed in Fig. 3.12a, the lower limit at which disturbance due to white

a Five-grade scale

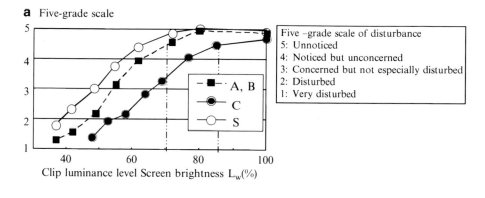

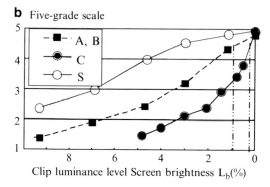

Five –grade scale of disturbance
5: Unnoticed
4: Noticed but unconcerned
3: Concerned but not especially disturbed
2: Disturbed
1: Very disturbed

Clip luminance level Screen brightness L_w(%)

b Five-grade scale

Clip luminance level Screen brightness L_b(%)

Fig. 3.12 Degree of disturbance due to white/black clips

clip is detected is a luminance level of about 70% in A and B and about 85% in C. In the case of disturbance due to black clip, it is about 1% in A and B and about 0.2% in C, as shown in Fig. 3.12b. It transpires that image impairment is larger in the case of differences in black than in white clip level.

Next, the same experiment was performed for A and B by reducing the size of geometric pattern blocks to a viewing angle of 10 arcmin. The resulting assessments are plotted as S in Fig. 3.12. It is apparent that the disturbance increases with the area of each block. In effect, the disturbance becomes worse as the area of each clipped domain expands, even if the total clipped area within the image remains the same.

3.2.3 Experimental Assessment Using Natural Pictures

The same experiment was also performed by using natural pictures. Two images were used, one of Football and the other of Autumn Tints, both selected from the stereoscopic standard charts [5] because they exhibited a relatively large amount of

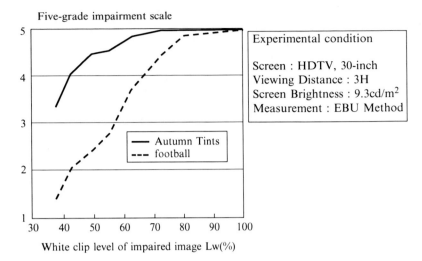

Fig. 3.13 Results of subjective evaluation for the degree of disturbance by white level clip with natural pictures

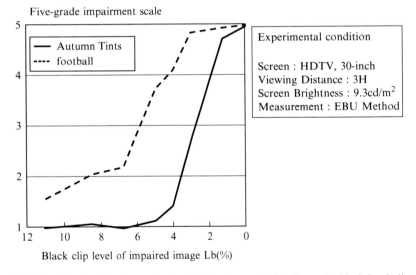

Fig. 3.14 Results of subjective evaluation for the degree of disturbance by black level clip with natural pictures

disturbance due to white and black clips. The parallax between left and right images was made zero (0) to avert any influence from Crosstalk in the evaluation. Figures 3.13 and 3.14 illustrate the results of assessments performed by eight individuals.

3.2.4   Considerations

The result of experiments using geometric images demonstrated that the disturbance was larger in Case C than in Cases A or B. It can be said, therefore, that clip level differences between left and right images generated during shooting produce greater, not lesser, disturbance when the image is displayed after compensating to adjust the clipped peak luminance. This may be because the adjustment of the clipped peak luminance causes further luminance level differences to arise between the left and right images in the middle luminance range, thereby increasing the overall sensitivity to disturbance. It would also seem possible to provide compensation by adjusting the clip levels between left and right images by clipping the picture on the unclipped side. That, however, would cause 'white crush' or loss of shadow detail across the entire image. In view of such considerations, it is necessary to devote the fullest attention to adjusting the differences between the non-linear characteristics, luminance clip in particular, of the left and right images of the camera.

Comparing S for Cases A and B in Fig. 3.12, we can see that the amount of disturbance differs according to the size of the pattern in both black and white clip levels. A significant difference was also observed in the distribution analysis. The amount of disturbance increases in both cases as the area of each block expands. In effect, the disturbance due to luminance clip increases as the area of each clipped domain expands even if the total area of the clipped parts within the image remains the same.

It is apparent from Fig. 3.13, showing the results of the assessments using natural pictures, that Football, with a higher detection limit of approximately 75%, entailed greater disturbance due to white clip than Autumn Tints. Regarding the histogram of luminance distribution for each picture design, Football had a higher percentage of luminance near the white clip, and this was one reason for the higher level of disturbance overall. Regarding the two picture types, the domains with a luminance level of 75% or higher are shown in black in Fig. 3.15. Compared with Autumn Tints, Football has more continuous domains with luminance levels above 75%, making each such area bigger. The clipping of larger continuous domains may be another reason for the higher level of disturbance. By contrast, Autumn Tints, where the luminance distribution is biased toward the black level side, exhibits greater disturbance in Fig. 3.14. This suggests that the amount of disturbance and disturbance curves differ widely in natural pictures due to biases in the areas of clipped domains and luminance distribution histograms arising from each picture design.

3.3   Crosstalk

3.3.1   Introduction

Most current binocular stereoscopic display systems, excluding head-mounted displays and a few others, occasionally produce Crosstalk disturbance in which the left

a

b

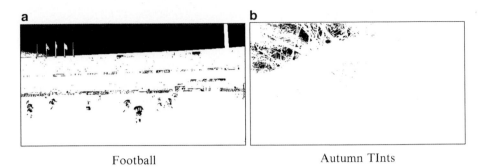

Football   Autumn TInts

Fig. 3.15 The domains of above 75% of luminance level of pictures used in the experiment

Fig. 3.16 Model of stereoscopic display with Crosstalk

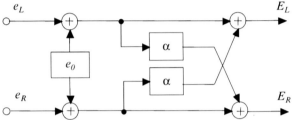

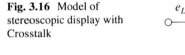

or right image is observed by the other eye. This not only impairs the quality of the stereoscopic image but also causes stress and tiredness for the viewer. It is, therefore, an important factor when seeking to develop a stereoscopic display system that is easy to watch to know the relation between the amount of Crosstalk and degree of disturbance.

Here, the lower limit for the detection of Crosstalk with regard to the contrast and parallax of the screen was investigated using geometric patterns. The disturbance characteristics due to Crosstalk were evaluated by subjective experiments using natural pictures and the results compared with those for ghost interference, a similar disturbance found in current television systems.

3.3.2   Generative Model of Crosstalk

Consider a binocular stereoscopic display system model with Crosstalk as shown in Fig. 3.16. The left and right image signal input corresponds to the luminance levels e_L and e_R, respectively, on the screen. The image here is displayed by adding setup level e_0 (the luminance level of the screen when the input signal is zero). The actual luminance levels observed by the left and right eyes become E_L and E_R, respectively, if any Crosstalk occurs.

In order to calculate α, the amount of Crosstalk with this display, we may first input a white level signal to the left side and black level signal to the right side. The luminance levels observed in such a case are described as follows when $E_L = W$ and $E_R = K$:

$$E_L = W = e_L + e_0 + \alpha e_0 \tag{3.6}$$

$$E_R = K = e_0 + \alpha(e_L + e_0) \tag{3.7}$$

Next, we input black level signals to both sides. Now, the observed luminance level is described as follows when E_L and E_R are E_0:

$$E_R = E_L = E_0 = e_0 + \alpha e_0 \tag{3.8}$$

Eliminating e_L and e_0 from Eqs. (6), (7) and (8), the amount of Crosstalk (α) can be described as follows:

$$\alpha = \frac{K - E_0}{W - E_0} \tag{3.9}$$

3.3.3 Measurement of Crosstalk Detection Limit

The lower limit for detecting Crosstalk was first investigated under the conditions in which it is most noticeable. The pattern of white-level marks on a black-level background shown in Fig. 3.17 was used in the experiment. Each mark is 100 arcmin high and 5 arcmin across. When the luminance of a black-level background is L, the luminance of desired image W, the amount of leakage from the other eye K, and amount of Crosstalk α are described by Eq. (9), and the image's contrast ratio C by Eq. (10), as follows:

$$C = \frac{W - E_0}{E_0} \tag{3.10}$$

Here, the lower limit for detecting Crosstalk was measured by varying the contrast ration C and parallax of the marks d. The binocular stereoscopic display system shown in Fig. 3.18 was used in this experiment, comprised of 20-inch CRTs and a mirror to establish a Crosstalk-free condition. Four people took the test at a viewing distance of 4 H. Note that K, E_0 and W are actual values measured by a spot-type luminance meter, and the peak luminance W is approximately 80 cd/m².

The results are shown in Fig. 3.19. Obviously, the detection limit becomes more difficult to determine with increases in screen contrast and parallax d. Although these results are somewhat different from the visibility threshold obtained by Pastool [6] a similar reliance on contrast and binocular parallax is shown in the

Fig. 3.17 Images used in evaluation

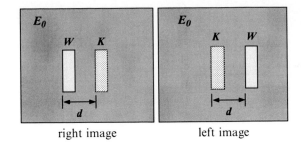

right image   left image

Fig. 3.18 Stereoscopic display system using mirrors

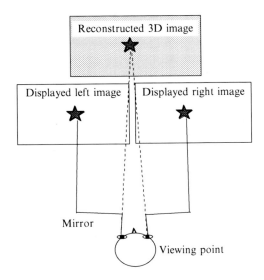

Fig. 3.19 Results of visibility thresholds for Crosstalk

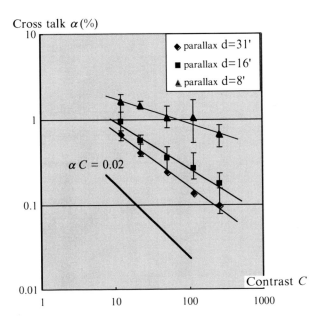

results. The graph also tells us that it can be approximated by a straight line for the respective parallax values within the range of measurement used this time, by describing α and C in logarithmic form.

Assuming that the influence of high luminance W in the screen can be ignored, Weber's law can be used to identify the leakage from the contralateral K for background luminance E_0 when the parallax is sufficiently large.

$$\text{Weber fraction}: \frac{K - E_0}{E_0} = \text{constant} \tag{3.11}$$

Here, the Weber fraction is about 0.02 [7] in the typical range of brightness found in everyday life. Equation (12) can therefore be derived from (9), (10), and (11) as follows:

$$\alpha \cdot C = \frac{K - E_0}{W - E_0} \cdot \frac{W - E_0}{E_0} = \frac{K - E_0}{E_0} = 0.02 \tag{3.12}$$

The graph is, therefore, approximated by the straight line described by Eq. (12) when the parallax is sufficiently large. In practice, however, there is also influence from high luminance W within the field of view when the parallax is big enough, and the graph is nearly saturated when parallax exceeds approximately 30 min. In the case of a display system at a contrast ratio of 100:1, for example, the lower detection limit for Crosstalk will be about 0.2%, according to the graph. In other words, the amount of Crosstalk needs to be below 0.2% in this case. The lower detection limit for Crosstalk in normal pictures could in fact be lower than this because the measurements were made under the severest image conditions taking Crosstalk into account.

3.3.4 Parallax and Crosstalk

Next, the relation between parallax and Crosstalk disturbance was verified using normal pictures.

Figure 3.20 shows the results of subjective evaluations using the stereoscopic image of Opera to discover the influences of Crosstalk disturbance when left and right images were shifted by 12 min or 24 min. The graph illustrates of the averages of evaluations performed in duplicate by six individuals.

The score for the standard picture is below 4.5 in the graph. This may be because the Crosstalk of the experimental system affected the evaluations near the lower detection limit, as we used a projected polarized stereoscopic display system in this experiment.

It is apparent from the results that larger horizontal shifts of left and right images cause a bigger impact from Crosstalk and increase the amount of disturbance even when the amount of Crosstalk is the same. In the case of Crosstalk of 10%, it is not

Fig. 3.20 Results of subjective experiments for variation of horizontal shifts

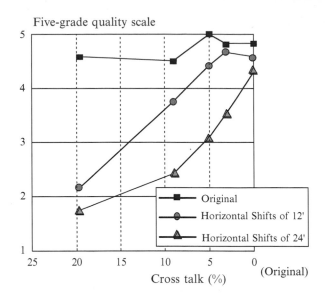

detected in the original image but is noticed when shifting the images 12 min horizontally and becomes intolerable with a horizontal shift of 24 min. In practice, the left and right images are sometimes displayed after being shifted horizontally to enhance the stereoscopic effect of the picture. It is therefore necessary for the operator to be aware that disturbance due to Crosstalk may increase in such cases.

3.3.5   Evaluation of the Crosstalk Disturbance Using Natural Pictures and Comparison with Ghost Interference in 2-D Images

The lower detection limit for Crosstalk was examined in Sect. 3.3.3; but, practically speaking, it is difficult to keep the amount of Crosstalk below 0.2% using current technologies. Disturbance due to Crosstalk was, therefore, evaluated in a subjective experiment that used basic natural stereoscopic pictures [8] to investigate the limits for Crosstalk in actual use.

In this experiment, a mock condition was produced by adding opposing side images to the left and right images. The gamma of the display was also taken into account, so that the Crosstalk could be handled as the amount of observed light rather than in terms of the electrical signal. The system shown in Fig. 3.18 was used for the experiment and three still pictures from the stereoscopic standard charts were chosen for the evaluation. The experimental conditions are outlined in Table 3.9.

Table 3.9 Experimental conditions

Measurement	EBU method
Data analysis	Three-way analysis of variance
Evaluated images	Three types (Football, City, Autumn Leaves)
Contrast	100:1
Screen size	20 inch
Image quality	High-definition
Screen brightness	White level 84 cd/m^2
Viewing distance	3.8 H
Lighting	Complete darkness
Length of presentation	10 s for both standard and test pictures
Evaluators	Seven individuals

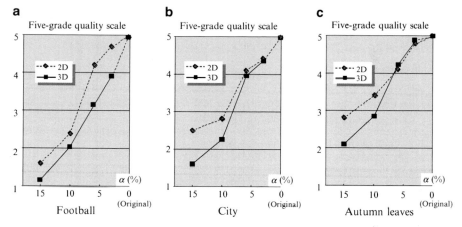

Fig. 3.21 Results of subjective evaluation for the degree of disturbance by Crosstalk

Figure 3.21 shows the results of evaluations performed by seven people. The limit of tolerance for Crosstalk disturbance (evaluation value: 3.5) is from about 5–10% for the pictures used in this experiment, but this depends on the pictorial design. The lower limit of detection (evaluation value: 4.5) is from about 1–2% for Football, in which the most Crosstalk was observed among the pictures used in this experiment. As far as Crosstalk is concerned, simply satisfying the bottom acceptable margin is not enough when the aim is to provide 3-D images of superior quality to viewers; the goal must be to bring it below 1–2%, thinking in terms of the lower detection limit for normal pictures.

Crosstalk disturbance seen with the 3-D image display was compared with ghost interference seen with 2-D images, by presenting the image from only one side to make it into a 2-D picture. The scores were lower for 3-D than for 2-D in general, indicating that 3-D pictures suffer a bigger impact for the same amount of disturbance.

3.3.6   Summary

The tests using normal pictures to measure the lower detection limit of Crosstalk indicated that Crosstalk should be kept below 1–2%. Considering the Crosstalk in terms of display method, polarization generates relatively little Crosstalk, but time-division using the LCD shutter and CRT causes more Crosstalk. Currently, Crosstalk with polarization is about 1% due to the influence of the screen, and it is about 5% with time-division due to the influence of persistence in the CRT and LCD shutter. Taking these results and the current state of Crosstalk into account, the disturbance due to Crosstalk is most noticeable in time-division types for some picture designs. Development of display systems with much less Crosstalk is anticipated in the future. Along with this endeavor to improve system performances, trials are also now being performed to reduce Crosstalk by means of signal processing [9, 10].

References

1. I.P. Howard and H. Kaneko, "Relative shear disparities and the perception of surface inclination", Vision Res., 34, No. 19, pp. 2505-2517 (1994).
2. H. Isono and M. Yasuda, "Conditions providing for field-sequential stereoscopic vision", J. Inst. TV Engrs. Japan, 41, No. 6, pp. 549-555 (1987).
3. C.W. Smith and A.A. Dumbreck, "3-D: the practical requirements", Television: J. Royal Television Society (Jan./Feb. 1988).
4. B. Choquiet et al, "3-D TV studies at CCITT", Proc. Of TAO International Symposium (Dec. 1993).
5. RECOMMENDATION ITU-R BT.1438 SUBJECTIVE ASSESSMENT OF STEREOSCOPIC TELEVISION PICTURES.
6. A. Pastool, "Human Factors in 3-D Imaging", HHI report'96.
7. K.B. Benson, "Television Engineering Handbook", McGraw-Hill, Inc.
8. M.Okui, et al., "Test materials for Evaluating Stereoscopic Television Systems", IBC99 pp. 565-570.
9. K. Hamada, et al., "A Field-Sequential Stereoscopic Display System with 42-in. HDTV DC-PDP", IDW'98, pp. 555-558.
10. J. Konrad, "Cancellation of image Crosstalk in time-sequential displays of stereoscopic video", IEEE Transactions of image processing, Jan. 1998.

Chapter 4
Psychological Factors and Parallax Distribution in the Case of 3-D HDTV Images

Abstract We have studied pictures obtained under different shooting and viewing conditions and many test images made for various production purposes, including the effective setting of objects to study the relationship between picture characteristics and the psychological effects they produce. For the purpose of this report, we chose to focus on the two key psychological factors of ease of viewing and sense of presence. Poor ease of viewing of scenes is regarded as one of the causative factors of visual fatigue. We also looked at the characteristic spatiotemporal distribution of parallax in 3-D images. While many earlier studies had focused on the accurate extraction of local parallax, we paid attention to the characteristics of screen parallax as a whole.

We then investigated the correlations between these psychological factors and the characteristics of parallax distribution in order to understand what constituted hard and easy images to watch with the goal of creating a new 3-D HDTV system that would minimize visual fatigue.

Keywords 3-D HDTV • Ease of viewing • Parallax distribution • Psychological effect • Psychological factor • Sense of presence • Stereoscopic HDTV • Visual comfort • Visual discomfort

4.1 Extracting Key Psychological Factors and the Patterns of Parallax Distribution

4.1.1 Introduction

In this section, we study the extraction of psychological factors from 3-D HDTV images with respect to parallax distribution patterns. We take the example here from studies on psychological factors produced by 3-D HDTV images [1] shot by a

H. Yamanoue et al., *Stereoscopic HDTV: Research at NHK Science and Technology Research Laboratories*, Signals and Communication Technology,
DOI 10.1007/978-4-431-54023-6_4, © Springer Japan 2012

3-D HDTV camera system, in which stereoscopic test materials with known shooting conditions and object positions were used.

We consider especially:

– Extraction of psychological factors from 3-D images (factor analysis)
– Analysis of the spatiotemporal distribution of parallax in pictures used in the evaluation (principal component analysis)
– Examination of correlations between the psychological factors and parallax distribution (correlation analysis)

4.1.2   Extraction of Psychological Factors from 3-D Images

We conducted a subjective evaluation using factor analyses to study the visual impressions made by 3-D images and their psychological effects.

4.1.2.1   Evaluated Images

For this subjective evaluation, we chose ten of the pictures used in a preliminary test. They were stereoscopic items for which shooting conditions and object positions were known, which had been used in the tests for the MPEG2 multi-view-profile (MVP), and which had been shot by a new compact 3-D HDTV camera using the zoom function. Some of the images are shown in Fig. 4.1. Each scene is a 15-s sequence and there was no rapid movement of either the object or the camera. Table 4.1 lists the titles, shooting conditions, main shooting range of the scene, and scene content.

4.1.2.2   Evaluation Test

We added their 2-D versions to these ten scenes, thus gathering subjective evalua-tions of a total of 20 images. These images were shown randomly on 120- and 70-inch screens by the 3-D HDTV system to observers wearing polarized glasses. Table 4.2 lists the experimental conditions. Each of these images was shown twice on both screen types. When assessing 2-D images, the left-eye image was displayed to both eyes. We made sure that the polarized glasses stayed fully in position and the subjects did not know whether to expect a 2-D or 3-D image. The subjects watched together in groups of two to four. The viewing distance was set at about 3H (H: screen height). The subjects consisted of 99 people, including both males and females, mostly in the 20- to 30-year-old age range, and their stereoscopic vision was checked before the test. On the basis of the preliminary test, 13 items were evaluated on a five-point scale, as shown in Table 4.3.

Fig. 4.1 Example images. Scene No. 3: flower pot. Scene No. 10: festival

4.1.2.3 Results

Factor analysis of the subjective assessments produced two factors with eigenvalues of more than 1. The contributions of these factors were 32% and 26%, respectively, making a cumulative contribution of 58%. Figure 4.2 shows the factor loading for each item in relation to these two factors. Of the 13 items, "looks real" had the highest factor loading and "easy to watch" was second. From these results, we call the "sense of presence" Factor 1, and "ease of viewing" is Factor 2 in this paper. Figure 4.3 is a scatter diagram of scores for images on the 120-inch screen with regard to these two factors.

Table 4.1 Evaluation images used in the test

No.	Scene	Focal length (mm)	Intersection point	Camera separation (mm)	Camera movement	Shooting range	Contents
1	Street organ	10	2.9 m	70	Dollying, panning	Near-middle distance	A girl and street organ
2	Aquarium	10	3.5 m	70	Fixed	Near	Fish in an aquarium
3	Flower pot	10	6.1 m	70	Dollying	Near-middle distance	A girl behind a flowerbed
4	Meal	12	Parallel	65	Fixed	Near	Meal at table
5	Market	12	Parallel	65	Fixed	Near-middle distance	A girl shopping in the market
6	Lion	40	Parallel	65	Almost fixed	Far	A lion on the prowl
7	Cheetah	40	Parallel	65	Almost fixed	Far	A cheetah roaming the savanna
8	Singer	12	Parallel	65	Fixed	Near	A singer in the studio
9	Bus stop	12	Parallel	65	Fixed	Near-far	People walking near a bus stop
10	Festival	12	Parallel	65	Panning	Near-middle distance	A portable shrine and swirling confetti

Table 4.2 Experimental conditions

Image types	10 (a total of 20 pictures in 2-D and 3-D)
Subjects	99 (males and females, from teenagers to those in their 60s)
Times viewed	Twice
Display system	3-D HDTV with polarized glasses
Screen size	120-inch, 70-inch
Peak brightness	21.4 cd/m² (120-inch)
	106 cd/m² (70-inch)
Viewing distance	Approximately 3H (H: screen height)
	4.5 m for 120-inch, 2.6 m for 70-inch

Table 4.3 Impressions for evaluation

Impressions	1. Distracted by the screen frame
	2. Easy to watch
	3. Looks like a miniature
	4. Image seems to go back a long way
	5. Looks real
	6. Tiring to watch
	7. Stands out from the screen
	8. Straining to eyes
	9. Depth seems natural
	10. Large
	11. Comfortable to watch
	12. Powerful
	13. Flat
Categories	5. Agree
	4. Somewhat agree
	3. Neither agree nor disagree
	2. Somewhat disagree
	1. Disagree

4.1.2.4 Considerations

Factor Loading

The 13 items used in this test can be divided roughly into two types, one to do with the subject's psychological and physiological impressions, including "feels compelling" and "strains the eyes," and the other to do with impressions of the image itself, such as "stands out from the screen" and "looks flat."

We turn first to the psychological and physiological impressions. In Fig. 4.2, "looks real" represents the Factor 1 axis and "easy to watch" that of Factor 2. We see that ease of viewing is little affected by Factor 1, and the sense of presence is

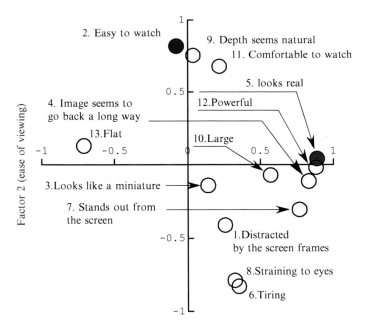

Factor 1 (sense of presence)

Fig. 4.2 Impact of the two major factors on subjective impressions

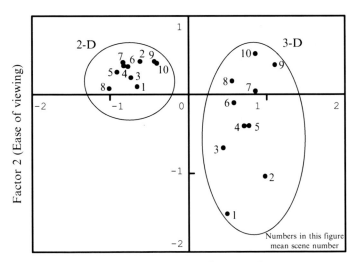

Factor 1 (Sense of presence)

Fig. 4.3 Factor scores of images in relation to Factors 1 and 2

also little affected by Factor 2. Conversely, impressions of tiredness and eye strain are both well removed from the two axes, suggesting that they are strongly affected by both Factors 1 and 2.

Turning then to impressions of the images themselves, the impressions of going back a long away and standing out from the screen are found on the plus side of the x-axis, indicating that 3-D images with a sense of depth or stereoscopic effect do produce a strong sense of presence. The sense of flatness is found at the minus end of the x-axis. This happened because half the pictures used in the test were 2-D. These 2-D images also scored poorly for sensation of depth. The impression of miniaturization, which refers to the puppet-theater effect and can spoil the sense of presence, had the smallest factor loading of all. This was because none of the 3-D images used in the test produced any strong puppet-theater effect.

On the y-axis, the naturalness of the depth sensation scores on the plus side, suggesting that 3-D images with natural depth perception (stereoscopic effects) are easy to watch. The question about distraction by the frame was designed to assess the frame's obstruction of binocular fusion when viewing 3-D images. This scored negatively because it makes the 3-D image hard to watch.

Factor 1: Sense of Presence

Figure 4.3 shows that 3-D images clearly have greater factor scores than 2-D images with regard to the sense of presence. To study this point in greater detail, we focused on the "looks real" impression, which has the largest factor loading for Factor 1, and analyzed its variance in terms of three factors: 2-D/3-D, scene, and screen size. The results show with a risk rate of 1% that the factorial effects and their interaction with the first two factors are significant. This variance analysis is shown in Table 4.4. Comparing the f-numbers of Table 4.4, 3-D images provide a much stronger sense of presence than 2-D images, though the degree seems dependent on scene content. On the other hand, no significant factorial effects were dependent on screen size. The results for brightness varied widely with screen size, as shown in Table 4.2, so we did not study any comparisons in this respect.

Factor 2: Ease of Viewing

We can see from Fig. 4.3 that all 2-D images scored nearly the same for this factor but there was a broad distribution for 3-D images according to the scene type. The score was very low for Scene 1, for example, but higher than for any 2-D image in the case of Scene 10. Ease of viewing is clearly very dependent on content. To study this in greater detail, we also analyzed the variance of the three factors of 2-D/3-D, scene, and screen size with regard to "Easy to watch", which has the largest factor loading for Factor 2. Table 4.5 shows the results.

Table 4.4 Analysis of variance for sense of presence

Factor	Sum of squared deviations	Degree of freedom	Mean square	f-number
A (2-D or 3-D)	2,445.59	1	2,445.59	3,349.54**
B (screen size)	0.21	1	0.21	0.29
C (scene)	110.29	9	12.25	16.78**
A×B	0.16	1	0.16	0.22
A×C	46.76	9	5.20	7.12**
B×C	5.40	9	0.60	0.82
A×B×C	2.98	9	0.33	0.45
Error	2,862.10	3,920	0.73	
Total	5473.49	3,959		

**1% significance

Table 4.5 Analysis of variance for ease of viewing

Factor	Sum of squared deviations	Degree of freedom	Mean square	f-number
A (2-D or 3-D)	337.75	1	337.75	579.79**
B (screen size)	0.43	1	0.43	0.75
C (scene)	350.70	9	38.97	66.89**
A×B	0.52	1	0.52	0.90
A×C	234.57	9	26.06	44.74**
B×C	10.41	9	1.16	1.99*
A×B×C	6.31	9	0.70	1.20
Error	2,283.54	3,920	0.58	
Total	3,224.24	3,959		

*5% significance; **1% significance

Comparing the f-numbers in Table 4.5, we find that significant factorial effects are derived not only from whether the image is 2-D or 3-D but also due to differences of scene and the interaction of these variables. Interaction is also detected between Factor B (screen size) and Factor C (scene). In the case of Scenes 4–10 (which were shot with parallel optical axes), the horizontal phase of the left and right images was adjusted according to the screen size, such that the infinite point during shooting will be at an infinite distance when displayed on the screen. Even identical scenes may generate differences of depth perception depending on screen size. As we see in Table 4.4, the factorial effects of screen size do not appear to affect ease of viewing.

4.1.3   Analysis of Parallax Distribution

We focused on the distribution of parallax in the examination of scene characteristics. As there was no rapid object or camera movements, we only analyzed the first frame in each test. In effect, we analyzed parallax on the basis of stationary pictures.

Fig. 4.4 Primary principal component loading by domain

Domain 1 $a_1 = 0.85$	Domain 2 $a_2 = 0.89$	Domain 3 $a_3 = 0.87$
Domain 4 $a_4 = 0.93$	Domain 5 $a_5 = 0.97$	Domain 6 $a_6 = 0.96$
Domain 7 $a_7 = 0.86$	Domain 8 $a_8 = 0.93$	Domain 9 $a_9 = 0.92$

Fig. 4.5 Secondary principal component loading of domain

Domain 1 $b_1 = 0.24$	Domain 2 $b_2 = 0.37$	Domain 3 $b_3 = 0.46$
Domain 4 $b_4 = -0.15$	Domain 5 $b_5 = 0.06$	Domain 6 $b_6 = 0.12$
Domain 7 $b_7 = -0.41$	Domain 8 $b_8 = -0.34$	Domain 9 $b_9 = -0.33$

4.1.3.1 Detecting Parallax

To detect parallax, we used block matching with a brightness of 16×16 pixels. In the case of 3-D HDTV images, we were able to obtain parallax distribution data from 64 vertical and 120 horizontal blocks. Of the ten scenes used in this test, those shot with parallel optical axes (Nos. 4–10) required horizontal phase adjustment of the left and right images according to the size of the screen. Parallax was therefore detected separately for these scenes on the 70- and 120-inch screens. The accuracy of parallax detection by block matching is largely dependent on the image type and precision drops where there is not much brightness variation. Errors were conspicuous, for example, in the background of Scene 8 (stage area behind a singer and band, where the brightness is flat and low), so this scene was not considered here. For the remaining nine scenes and two screen sizes, we obtained 15 sets of parallax distribution data. The data were then divided into nine domains to facilitate analysis. Such domains would generally be quarter-sized both vertically and laterally. As the main object typically occupied a large area in the screen's center, however, the domains were weighted in accordance with their positions, as shown in Figs. 4.4 and 4.5. Specifically, the screen was divided into the nine domains of the center (with domain 5 taking up 4/16th of the total screen area) and periphery. Peripheral domains 2, 4, 6, and 8 each occupied 2/16th and the 4 corners (1, 3, 7, and 9) 1/16th of the screen. The parallax distribution data within each domain were then averaged. Here, negative values indicate cross parallax that projects forward from the screen and positive values non-crossing parallax aligned toward the back of the screen. Principal component analysis applied to the parallax data in these nine domains showed that the data could be grouped into two principal components with a cumulative contribution factor of 92.5% (83.2% for the primary principal component and 9.3% for the secondary one). Figures 4.4 and 4.5 show the loading of these two principal components for each domain.

4.1.3.2 Results of Principal Component Analysis

The primary principal component score for each image, S_1 is expressed by the equation

$$S_1 = \sum_{i=1}^{9} a_i \times (\text{amount of parallax in each domain})$$

(a_i: see Fig. 4.4; i: domain number).

The coefficients of these domains (see Fig. 4.4) are all positive, falling into the narrow range of 0.85–0.97. This result indicates that, in the case of 3-D images, the greater the parallax in the positive direction (the farther toward the back), the higher the score for the primary principal component.

The secondary principal component score for each image, S_2 is expressed by the equation

$$S_2 = \sum_{i=1}^{9} b_i \times (\text{amount of parallax in each domain})$$

(b_i: see Fig. 4.5; i: domain number).

The coefficients of the domains for the secondary principal component (see Fig. 4.5) are positive for the top three domains (1, 2, 3) and negative for the bottom three (7, 8, 9). This indicates that, in the case of 3-D images, the farther the top is towards the back of the scene and the bottom toward the front of the scene, the higher the score for the secondary principal component.

4.1.3.3 Distribution of Parallax Between Domains (Dispersion)

Some 3-D images are very deep, ranging from near the viewer to an infinite distance, while others are very shallow. We added this element as the inter-domain dispersion of parallax data.

$$S_3 = \sum_{i=1}^{9} (P_i - \hat{P})^2$$

where

$$\hat{P} = \frac{\sum\limits_{i=1}^{9} P_i}{9}$$

4.1.4   Correlation Between Psychological Factors and Parallax Distribution

We analyzed the correlation between each image's primary and secondary principal component scores, the inter-domain dispersion obtained through the principal component analysis of parallax distribution, and the factors obtained by subjective evaluation (multiple regression analysis).

4.1.4.1   Correlation with Factor 1: Sense of Presence

No clear correlation was found between the sense of presence and primary principal component, secondary principal component, or parallax dispersion (the determinant coefficient of multiple regression was 0.4, indicating that the regression was not significant). In effect, no relationship was observed between the sense of presence of the 3-D image and the distribution of parallax vectors.

4.1.4.2   Correlation with Factor 2: Ease of Viewing

Clear correlations were found between the ease of viewing and primary principal component, secondary principal component, and parallax dispersion (the determinant coefficient of multiple regression was 0.85, indicating that the regression was significant). In effect, a relationship was observed between the ease of viewing of the 3-D image and the dispersion of parallax vectors. A regression formula was used to obtain:

$$\text{Ease of viewing factor} = 0.36 \times S_1 \left(\text{primary principal component score}\right)$$
$$+ 1.0 \times S_2 \left(\text{secondary principal component score}\right)$$
$$- 0.9 \times S_3 \left(\text{dispersion of parallax vectors}\right) + \text{constants}$$

Of the coefficient of this equation, those of the primary and secondary principal component scores are positive, and that of parallax dispersion is negative. It appears that 3-D images that are easy to watch score highly for the primary and secondary principal components and have little dispersion of parallax. In terms of the absolute values of these coefficients, the secondary principal component and parallax dispersion are more influential than the primary principal component. From these results, we conclude that a 3-D image is easier to watch if the parallax is shaped such that the bottom projects forward from the screen and the top is drawn back toward the rear. It is also significant that there is less irregularity of parallax between large areas. The 3-D images that are located toward the back of the frame are also easier to view than those in the front.

4.1.5 Summary

We examined ten scenes of stereoscopic test materials with known shooting conditions and object positions and compared them with corresponding 2-D images in terms of such factors as the sense of presence and ease of viewing as obtained by subjective evaluation. The results revealed that 3-D images provide a stronger sense of presence than 2-D images but the ease of viewing is largely dependent on the type of scene. In fact, some 3-D images were judged easier to watch than their 2-D counterparts. Next, to analyze scene content, we performed principal component analysis for the parallax distribution of each image and extracted characteristics from the scenes. Lastly, we performed multiple regression analysis to clarify the correlations between the characteristics of parallax distribution and sense of presence/ease of viewing. We found no clear relationship between parallax distribution and sense of presence, but there was a strong relationship between parallax distribution and ease of viewing. The multiple regression formula thus obtained was used to infer which 3-D images would be easy to watch. It remains, however, to extract the characteristics more precisely from the parallax distribution. This requires a larger number of domains and the study of images with more complex parallax distribution patterns. The results we have at present tell us that a 3-D image is easier to watch if the parallax is shaped such that the bottom projects forward from the screen and the top is drawn back toward the rear. Significantly, there is typically little irregularity of parallax between large areas. The 3-D images that are located toward the rear of a scene are also easier to watch than those in the front.

4.2 The Range of Parallax Distribution, and Its Application to the Analysis of Visual Comfort in Stereoscopic HDTV

4.2.1 Introduction

Stereoscopic HDTV based on conventional HDTV technology has good picture quality, and many people can enjoy these programs at the same time by using a large screen. A variety of equipment has been developed in this field, including cameras [2] and displays, and has been used in the production of many stereoscopic HDTV programs. Among recent examples are those that covered the Summer Olympic

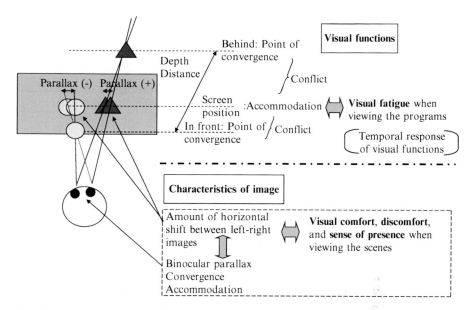

Fig. 4.6 Background of our study

Games in Sydney, Australia and the Winter Olympic Games in Salt Lake City (Utah, USA). The Winter Olympic Games in Nagano provided an opportunity to test the transmission of stereoscopic HDTV programs on two left–right channels via a satellite. NHK also experimented with IP transmission via a gigabit network. As these attempts show, stereoscopic HDTV is now very close to practical application in closed circuits in terms of equipment, systems, and program production techniques. However, some problems have been identified through these activities, such as that viewers may experience visual fatigue while watching or after watching stereoscopic HDTV programs and some scenes in these programs are rather hard to see depending on how they are presented.

Differences of electronic characteristics between left and right images, crosstalk, and excessive binocular parallax are likely causes of viewing discomfort and visual fatigue. As for visual functions, the conflict between the eye's accommodation and convergence, which is characteristic of the stereoscopic display system, is thought to be a major factor contributing to these problems. The conflict of convergence and accommodation depends on where the image appears in relation to the position of the screen, and this location of the image is influenced by the amount of horizontal shift between left and right images. In this paper, we call this horizontal shift "parallax". On the other hand, parallax distributions influence visual comfort and sense of presence, too. It should be noted that scenes in a stereoscopic HDTV program flow in a continuous manner along the time axis and that each scene contains some movements. It is therefore essential that any research on stereoscopic television viewing deals with the time response of visual functions and visual comfort (or discomfort) influenced by the temporal change of parallax. Figure 4.6 illustrates the background and objectives of our study, based on the above circumstances.

Our study attempts to combine the results of these previous investigations to produce stereoscopic HDTV services that are less straining on the eyes and more attractive to viewers.

In this paper, we report on topics shown within the broken-line box in Fig. 4.6. First, to measure parallaxes from stereoscopic HDTV pictures, we propose a method based on phase correlation [3]. Two types of threshold processing were applied to the phase-correlation surface in order to minimize measurement errors influenced by picture patterns. Second, we conducted a subjective evaluation test regarding visual comfort and the sense of presence, using 48 still images. We applied the proposed method to these test pictures, measured their parallaxes and then compared the results with those of the subjective evaluation test. The comparison results clarified the correlation between parallax distribution and visual comfort and between parallax distribution and the sense of presence.

4.2.2   Measurement of Parallax

4.2.2.1   Measurement Conditions

With the current technology, it is very difficult to measure parallaxes by each pixel or by each small block in a stereoscopic image. When measuring parallaxes for each object in a frame, it is also difficult to extract each object for measurement. The primary purpose of parallax measurement is therefore limited to studying the relationship between the characteristics of parallax distribution and visual comfort, and the relationship between the characteristics of parallax distribution and the sense of presence when watching the scenes. We therefore consider that we need to obtain only statistical data concerned with parallax. In other words, it is not necessary to know pixel-by-pixel parallax or to measure the parallax of every object. The following practical characteristics of parallax are considered:

- Range of parallax distribution across the display area
- Dispersion of parallax within the display area
- Average of parallax within the display area
- Temporal changes in the above figures

The following should be considered when measuring the above characteristics:

- The parallaxes of as many objects as possible within the picture should be measured. However, it is not necessary to match between parallaxes and the objects for which they are measured. Pixel-by-pixel measurement of parallax in left–right images is not necessary.
- It is important to avoid error detection of parallax when measuring the range of parallax distribution, though it is undesirable to miss necessary parallaxes.
- Sufficient calculation capacity is required so that the whole parallax of all frames of a relatively long program can be measured as quickly as possible.

4.2.2.2   Measurement Method

We propose a parallax measurement method based on phase correlation in consideration of the above points, which offers the following advantages:

- Use of a high-speed FFT algorithm
- Resistance to noise (fewer error detections)
- Independent of level differences between left and right images

The greatest merit of using a phase correlation method for this parallax measurement is that, unlike TV standards conversion [3] or high-efficiency coding, it is not necessary to reassign the results of parallax measurement to each of the segmented blocks or pixels. Namely, for our purpose, the parallaxes measured for a relatively large block are used as a representative figure for an image in the block.

When measuring the parallax by this phase-correlation method, error detection tends to occur under the following circumstances:

(1) When the picture inside the block has little texture or when parallax is continuously changing, the noise level increases on the corresponding phase-correlation surface and remarkable peaks of the surface do not appear as a result.
(2) In a stereoscopic picture, areas where left and right images do not match appear on the left and right sides of the object. This causes noise around the peaks that correspond to the object on the phase-correlation surface.
(3) Influences from the edges of the measured block cause spurious zero and noise on the phase-correlation surface.

To deal with problem (3), we normally use a window function for the picture inside the block, reducing the signal level to zero at the block's edges. This paper proposes a new way to reduce the erroneous findings noted in (1) and (2).

4.2.2.3   Parallax Measurement System

Figure 4.7 is a block diagram of the proposed parallax measurement system. The system consists of three parts: the pre-processing part, the part for calculating the phase-correlation surface, and the part for processing measured parallax. As shown in Fig. 4.8, left and right images are divided into measured blocks in the pre-processing part. The block size is 512 pixels × 512 lines, based on the fact that the largest parallax of HDTV programs produced is about 150 pixels in size. This block size is fairly large. As Fig. 4.8 shows, these blocks overlap rather extensively, so that we can measure the parallax of as many objects as possible. As a single block may have several objects, meaning there are several parallaxes inside it, we decided to measure up to three representative parallaxes within the block.

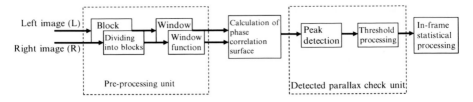

Fig. 4.7 Block diagram of parallax measurement

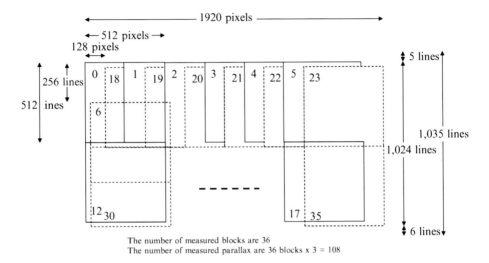

The number of measured blocks are 36
The number of measured parallax are 36 blocks x 3 = 108

Fig. 4.8 Measured blocks in a picture

This arrangement of blocks means that there are 36 blocks within a single picture, and that up to 108 parallaxes (36 blocks × 3 parallaxes) can be measured. The window function was chosen to eliminate the influence of block edges, as follows:

$$\left(1 - \cos^2\left(mn/512\right)\right)\left(1 - \cos^2\left(nn/512\right)\right) \quad 0 \le m, n < 512 \; m, n : \text{integers}$$

where, m and n represent the number of horizontal and vertical directions, respectively. Then, the value of the center of the block is 1 and the values of the block edges are 0.

The phase-correlation surface $z(x, y)$ is calculated as follows:

When $f_L(x, y)$ represents left images and $f_R(x, y)$ represent right images, the Fourier transformations are:

$$F_L\left(X, Y\right) = F\left\{f_L\left(x, y\right)\right\}$$

$$F_R\left(X, Y\right) = F\left\{f_R\left(x, y\right)\right\}$$

where, F{ } means Fourier transformation.

Now, Z(X, Y) is defined as follows:

$$Z\left(X,\ Y\right)= F_{L}\left(X,\ Y\right)\times F_{R}\left(X,Y\right)*/\left|F_{L}\left(X,Y\right)\times F_{R}\left(X,Y\right)*\right|$$

where, * means common complex number.

Then, the phase-correlation surface z(x, y) is

$$z\left(x,\ y\right)= F^{-1}\left\{Z\left(X,\ Y\right)\right\}$$

where, F^{-1} { } means reverse Fourier transformation.

In parallax measurement by the phase-correlation method, peaks on the phase-correlation surface are detected, and the coordinates of these peaks represent the parallaxes. To reduce the errors shown in Sect. 4.2.2.2 (1) and (2) when detecting parallaxes from the phase-correlation surface, we introduce two threshold-processing routes for the phase-correlation surface.

4.2.2.4   Method to Reduce Error Detection and Its Effectiveness

Threshold Processing 1

Figure 4.9 shows two threshold processing routes to reduce detected errors. Threshold Processing Route 1 is for reducing detected errors attributable to picture patterns [Sect. 4.2.2.2 (1)]. It estimates whether the maximum peak of the phase-correlation surface reflects the correct parallax of the object or whether the peak is caused by noise. The principle is as follow:

For a phase-correlation surface z(m, n): $0 \leq m$, $n \leq N\text{-}1$ (m and n are integers, N is 512), the maximum value of z(m, n) is defined as: Zmax = max[z(m, n)] ($0 \leq m$, $n \leq N\text{-}1$).

Where, the number of coordinates (m,n) is defined as k when satisfying the condition:

$$z\left(m,\ n\right)\geq \alpha \times \text{Zmax}\ \left(0m,\ nN-1\right).$$

Then S(α) is defined as:

$$S(\alpha) = \left(k \times 100\right)/ N^{2}.$$

S(α) now shows the characteristics of the phase-correlation surface. Figure 4.10 shows examples of representative pictures inside the measured block

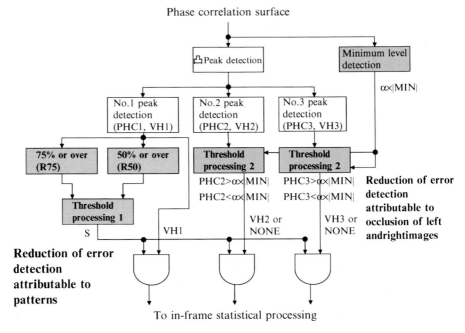

Fig. 4.9 Threshold processing

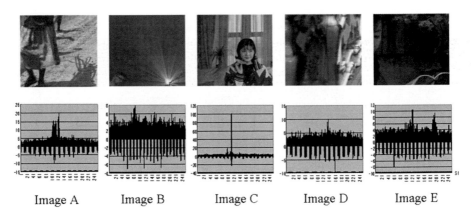

Fig. 4.10 Examples of representative images and phase correlation surfaces inside the measured block

and corresponding phase-correlation surfaces. These phase-correlation surfaces show how 3-D phase-correlation surface graphs look when they are viewed from a direction parallel to the vertical axis. On the horizontal axis, 129 represents parallax 0, 1 represents parallax 128, and 256 represents parallax -127. The unit of these values is a pixel of picture width. Figure 4.11 shows the values of $S(\alpha)$ of these pictures shown in Fig. 4.10. We can see, for instance, that $S(\alpha)$ moves to Y when a

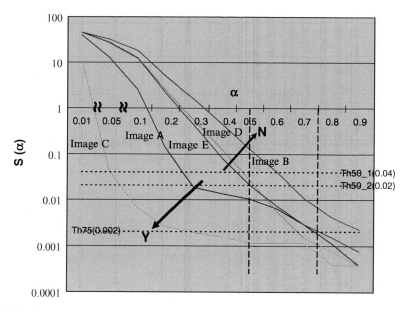

Fig. 4.11 S(α) against Fig. 4.10

picture within the block has obvious textures (such as Images A and C) and there are obvious peaks on the surface, and that S(α) moves to N when a picture within the block does not have obvious textures (such as Image B) and there are no obvious peaks on the surface. Thus, S(α) can represent the characteristics of picture patterns and the phase-correlation surfaces.

Another example is shown in Fig. 4.12. Figure 4.13 shows the S(α) of all the measured blocks within Fig. 4.12. From Fig. 4.13, we can predict that the phase-correlation surface in blocks 17, 34, and 35 contains a lot of noise, and that other blocks have obvious peaks. Blocks 17, 34, and 35 are those without any character-istic patterns, located in the lower-right corner of the picture in Fig. 4.12. Figure 4.14a shows the phase-correlation surface of Block 17, and Fig. 4.14b with a sharp peak is that of Block 8 (a square in the center of the picture). From the example of the Fig. 4.12 picture, we can see that errors caused by the picture in the measured blocks which have little obvious texture are estimated by using S(α) and there is a possibility of decreasing these errors. The key here is the inclusion of S(α), which shows the characteristics of the phase-correlation surface. For simplicity in execut-ing this processing, we use only two figures, α =0.75 and α =0.5, for S(α).

First, we set the flowing conditions:

– The total number of pixels within the block is MN (the block size is 512 pixels×512 lines, and so MN=512×512).
– The maximum peak value on the phase-correlation surface (No. 1 peak value) is PHC1.

Fig. 4.12 Example of test images

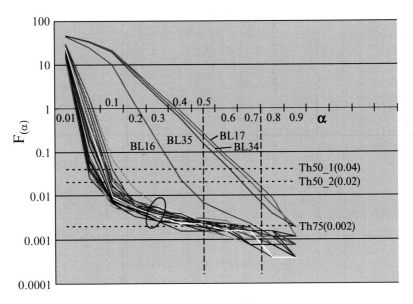

Fig. 4.13 $S(\alpha)$ against all measured blocks within Fig. 4.12

– The number of pixels at which the value on the phase-correlation surface is $0.5 \times PHC1$ or greater, is R50.
– The number of pixels at which the value on the phase-correlation surface is $0.75 \times PHC1$ or greater, is R75.

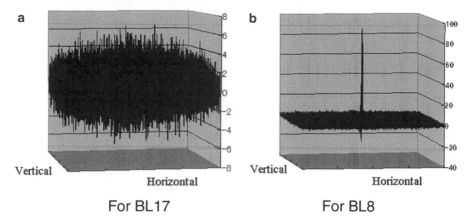

Fig. 4.14 Phase correlation surfaces for a picture in the measured blocks shown in Fig. 4.12.
(**a**) Correlation surface for Block 17. (**b**) Correlation surface for Block 8

We then define the following:

(a) $(R50 \times 100)/MN < 0.02$	$\rightarrow F = 0$
(b) $(0.02 \leq (R50 \times 100)/MN < 0.04) \cap ((R75 \times 100)/MN < 0.002)$	$\rightarrow F = 99$
(c) $(0.02 \leq (R50 \times 100)/MN < 0.04) \cap ((R75 \times 100)/MN < 0.002)$	$\rightarrow F = 0$
(d) $(R50 \times 100)/MN < 0.04$	$\rightarrow F = 99$

where $F = 0$ means that the picture pattern in the block is appropriate for parallax measurement by the phase correlation method, that is, the picture pattern has obvious textures, while $F = 99$ means just the opposite, that is, not appropriate for this kind of measurement [Sect. 4.2.2.2 (1)]. The above threshold values are plotted in Figs. 4.11 and 4.13. From Fig. 4.11, we can see that the Images A and C are the above-mentioned Case (a), Image B is Case (d), Image D is Case (b), and Image E is Case (c). Figure 4.13 shows that the pictures in Blocks 17, 34, and 35 in Fig. 4.12 are Case (d), and the other blocks are Case (a). Thus, Threshold Processing Route 1 works well with these pictures.

Threshold Processing 2

Threshold Processing Route 2, on the other hand, is designed to deal with the problem described in Sect. 4.2.2.2 (2). The purpose is to evaluate, when the phase-correlation surface in the block has several peaks, whether the second and third peaks correspond to several objects with a multiple number of parallaxes or whether these peaks are caused by the noise around the maximum attributed to the mismatch of left and right images. The procedure for Threshold Processing Route 2 is as follows:

We set the following conditions:

- The maximum peak value (No. 1 peak value) on the phase-correlation surface is expressed by PHC1 and its corresponding parallax by VH1.
- The No. 2 peak value on the phase-correlation surface is expressed by PHC2 and its corresponding parallax by VH2.
- The No. 3 peak value on the phase-correlation surface is expressed by PHC3 and its corresponding parallax by VH3.
- The minimum value on the phase-correlation surface is expressed by MIN.

We then define the following:

- If PHC2 is larger than $\alpha \times |\text{MIN}|$, then VH2 is detected as the correct parallax corresponding to an object within the block.
- If PHC3 is larger than $\alpha \times |\text{MIN}|$, then VH3 is detected as the correct parallax corresponding to an object within the block.
- In other cases, VH2 and VH3 will be ignored as noise. We assigned a value of 1.5 to α in this study.

Next, an example of the effects of Threshold Processing Route 2 is shown. We first counted parallaxes pixel by pixel from the left–right images of the two kinds of stereoscopic picture show in Fig. 4.15, and they are located at the measuring points. The numerical values of Fig. 4.15 show the results. We counted (a) at 41 points, counting 26 parallaxes, and (b) at 28 points, counting 19 parallaxes. On the other hand, Fig. 4.16 shows the results of the Route 2 processing in the proposed measuring method. For the Route 1 processing, $S = 0$ for all the measured blocks of the two images in Fig. 4.15. In Fig. 4.16, the 36 measurement blocks on the picture are represented by the horizontal axis and values on the phase-correlation surface by the vertical axis. Shown in the figure are three peaks (PHC1, 2, 3) inside each of the measured blocks and threshold values, that is, $\alpha \times |\text{MIN}|$ in each block by Route 2 processing. Circles in the graphs correspond to clear error parallaxes when compared with Fig. 4.15; the value of error-detected parallax is given beneath each circle. As has been show, because all circles are under the threshold values, these errors can be eliminated by the Route 2 processing. In summary, Threshold Processing Route 2 allows us to detect all the parallaxes of the main objects inside the two images shown in Fig. 4.15 while avoiding clearly erroneous parallaxes.

4.2.2.5   Parallax Measurement by Block Matching

To provide a comparison with the proposed system, we examine parallax measurement using a block-matching method for three block sizes: 16 pixels × 16 lines, 64 pixels × 64 lines, and 128 pixels × 128 lines. The search area was between −64 pixels and +63 pixels horizontally and vertically. MS expresses the yardstick of mismatching between two blocks, which is the absolute value of difference of luminance signal levels between corresponding pixels in two blocks on left and right images.

Fig. 4.15 Parallax of representative objects in test materials (each photo represents a *left* image). (**a**) Test material "In My Room". (**b**) Test material "City Scenery 2"

$$MS = \sum \sum \left| f_L(x, y) - f_R(x, y) \right|$$

Figure 4.17 shows the detection results for the image shown in Fig. 4.15a. These results show the appearance, when viewed from above, of the contours representing the amount of detected horizontal parallaxes at each point of the image. They show in gradation the amounts of parallax, as shown on the right of (a). As Fig. 4.17a indicates, the image which represents the amounts of parallax is severely fragmented in a block of size 16 pixels × 16 lines. This means that error

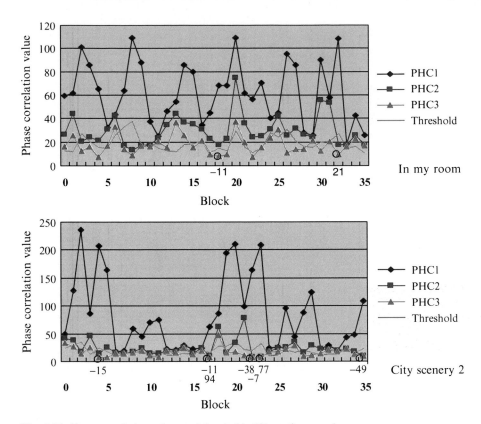

Fig. 4.16 Phase correlation values and threshold of Route 2 processing

detection is frequent. In an actual measurement using this 16-pixel, 16-line block, the maximum parallax always tends to become +63 pixels and the minimum parallax −64 pixels because of noise-induced errors. This is undesirable when measuring the range of parallax distribution.

To avoid this problem, we need to increase the size of each block to 64 pixels × 64 lines or 128 pixels × 128 lines and devise some measure to deal with detected errors. For each block size, we were able to measure the parallax of the main objects show in Fig. 4.15a by block matching. As seen in Fig. 4.17, by block matching, when compared with phase correlation, there exists a positional correlation between the detected parallax and the image; the smaller the block, the more precise the correlation. However, this positional correlation is not necessary for our study because we wish to obtain the characteristics, especially the range of parallax distribution within a picture, and so it is important to reduce the detected error. To do this, we need a fairly large block for block matching and a wide search range, requiring an extremely large calculation power. Since we intend to use a personal computer to process the characteristic parallax values of a relatively long program, phase correlation is preferred over block matching because high-speed FFT computation algorithms can be used.

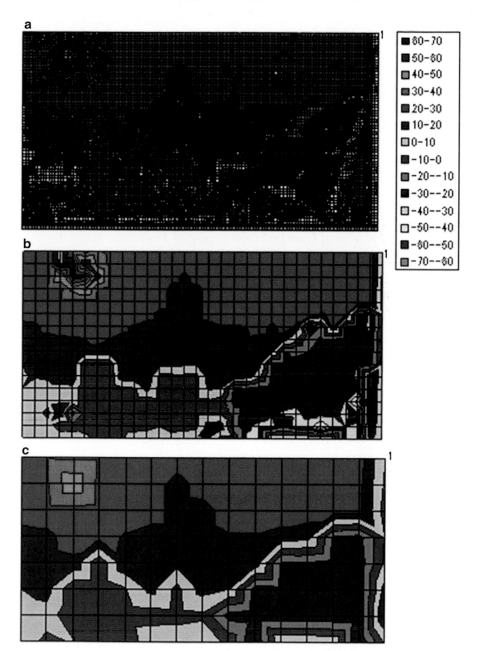

Fig. 4.17 Results of detection for parallax by block-matching method. (**a**) Block size 16 pixels × 16 lines. (**b**) Block size 64 pixels × 64 lines. (**c**) Block size 128 pixels × 128 lines

Table 4.6 Conditions of subjective evaluation test

Images used in test	48 still images (including a standard pattern)
Subject	24 adult males and females (not expert)
Repeat test	10-s viewing of 2-D image (for reference), followed by 10-s viewing of 3-D images (for evaluation)
Display system	3-D HDTV using polarizing glasses
Screen size	90 inch
Viewing distance	About 3H (3.33 m)
Peak brightness	15 cd/m²
Method of evaluation	Relative evaluation on a scale of 7, based on 2-D image

4.2.3   Parallax Distribution and Subjective Evaluation Test for Visual Comfort and Sense of Presence of Stereoscopic HDTV

4.2.3.1   Subjective Evaluation Test

We carried out a subjective evaluation test to assess the visual comfort and the sense of presence that people experience when watching a stereoscopic HDTV picture. Table 4.6 shows the experimental conditions. Images used for evaluation were scenes taken from stereoscopic HDTV test materials [4] and programs produced in the past. These images, numbering 48 in all, were all still pictures. The test subjects were shown stereoscopic HDTV (2-D) presentations of these pictures and conventional HDTV versions (2-D) that served as reference images. For the 2-D presentation, the left images were shown to both eyes, because the subjects can evaluate both 3-D and 2-D images while wearing the polarizing glasses. The results were evaluated on a scale of 7.

4.2.3.2   Relationship Between Subjective Evaluation Test Results and Parallax Distribution

We used the method described in Sect. 4.2.2 to measure the following as the statistical data of parallax distribution across the screen on which the test image was shown: average parallax over the entire screen, absolute value of the average, range of distribution, maximum value (parallax of the farthest object from the viewer), minimum value (parallax of the nearest object from viewer), and level of dispersion. Parallax was expressed by the pixel; it is zero on the screen, a positive figure away from the screen, and a negative figure when projecting out toward the viewer. Under the viewing conditions given in Table 4.6, parallax of 1 pixel means that left and right images are about 1 mm apart from each other on the screen. Table 4.7 shows the correlation between the statistical data of parallax obtained by the method proposed in Sect. 4.2.2 and visual comfort. The correlation between the range of parallax distribution and visual comfort is highest in the table, and the value is |−0.86|.

Table 4.7 Correlation between the statistical data of parallax and visual comfort

Simple correlation	Visual comfort	Average	|Average|	Range	Min.	Max.	Distribution
Visual comfort	–	–	–	–	–	–	–
Average	0.46	–	–	–	–	–	–
|Average|	−0.32	0.19	–	–	–	–	–
Range	−0.86	−0.46	0.32	–	–	–	–
Min.	0.76	0.84	0.02	−0.83	–	–	–
Max.	−0.35	0.50	0.60	0.49	0.07	–	–
Distribution	−0.80	−0.34	0.40	0.90	−0.67	0.57	–

Table 4.8 Correlation between the statistical data of parallax and sense of presence

Simple correlation	Visual comfort	Average	|Average|	Range	Min.	Max.	Distribution
Visual comfort	–	–	–	–	–	–	–
Average	0.21	–	–	–	–	–	–
|Average|	0.24	0.97	–	–	–	–	–
Range	0.65	0.03	0.06	–	–	–	–
Min.	−0.11	0.84	0.82	−0.47	–	–	–
Max.	0.46	0.90	0.90	0.40	0.62	–	–
Distribution	0.49	0.06	0.06	0.88	−0.43	0.32	–

As for the sense of presence and these statistical data of parallax distribution, we examined all the evaluation images, but did not find significant correlations between them. We then limited our examination to 35 images whose 3-D presentations had been found to be more comfortable to watch than their 2-D versions, and searched for any significant correlation between them. The results, as shown in Table 4.8, show that the correlation between the sense of presence and the range of parallax distribution is the highest at 0.65. The range of parallax distribution thus affects both visual comfort and sense of presence.

4.2.3.3   Visual Comfort and Sense of Presence Depending on Parallax Distribution

Range of Parallax Distribution and Visual Comfort

As clarified in Sect. 4.2.3.2, visual comfort has the strongest correlation with the range of parallax distribution. Figure 4.18 shows the relationship between visual comfort and the range of parallax distribution of each evaluation image, with the reference level being at the screen position. The vertical axis shows the distance from the screen (unit: pixel) and the horizontal axis represents visual comfort. Each dot expresses the average parallax distribution of each image, and the up–down bars show the range of distribution. We can see that those comfortable-to-view pictures have a smaller range of distribution and the magnitude of this range is below about 60 pixels. Furthermore, the parallax of the comfortable-to-view pictures falls between −30 pixels and +65 pixels.

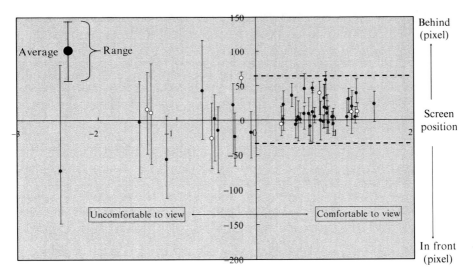

Fig. 4.18 Relationship between visual comfort and parallax distribution

Visual Comfort and Sense of Presence Against Average Parallax Distribution

To consider the relationship between average parallax distribution and visual comfort or the relationship between average parallax distribution and the sense of presence, we selected seven pictures (those indicated by reversed dots) from among the images shown in Fig. 4.18 and then moved to the horizontal positions of their left–right pictures by ±33 pixels and ±66 pixels. By shifting the average positions of parallax distribution in this way, we carried out an evaluation test on visual comfort and the sense of presence. A total of 20 people took part in the experiment, which was carried out under the same conditions as shown in Table 4.6. Figure 4.19 shows the results of this experiment on visual comfort. The data indicated by the reversed plotted line are those of the original picture without horizontal shift. The data show that the closer to the screen the average parallax distribution is, the more comfortable the picture is for the viewers to watch. The sense of presence, on the other hand, showed no such correlation.

4.2.4   Discussion

Under the same viewing conditions as the subjective evaluation test, pupil diameters were measured when subjects viewed the stereoscopic HDTV program "Waffen", and were found to vary from 2 to 5 mm. Assuming that the pupil diameter is 5 mm, the depth of field corresponding to the eye's focal depth is about 0.3 D (Diopter) [5]. Researchers measured the response in the eye's accommodation against changes in horizontal distance between left and right pictures during stereoscopic image

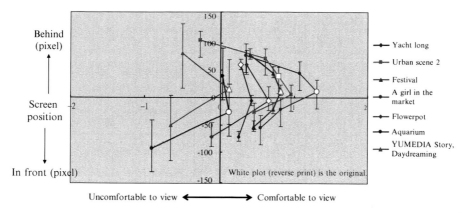

Fig. 4.19 Visual comfort when shifting horizontal distance between *left* and *right* images

observation under similar viewing conditions, and showed that the responsive range of accommodation is about 0.2–0.3 D [6]. On the other hand, as shown in Fig. 4.18, the range of parallax distribution when watching comfortably is about 60 pixels, and this range is about 0.3 D. We suggest that the depth of field corresponding to the eye's focal depth when the pupil is dilated may be related to visual comfort when viewing stereoscopic HDTV pictures.

4.2.5   Summary

This paper proposed a method for measuring the parallax of stereoscopic HDTV pictures. To reduce erroneous detection by the phase correlation method, we used two new threshold processing approaches. We also conducted a subjective evaluation test using 48 still pictures to study visual comfort and the sense of presence. The proposed method was used to measure the parallax of these images to clarify the relationship between the statistical data about parallax distribution and the results of the evaluation test. We found the following:

- The range of parallax distribution showed the strongest correlation with visual comfort. The range of the parallax distribution of pictures judged comfortable to view was almost 0.3 D, which is almost the same value as the depth of field corresponding to the eye's focal depth when the pupil is dilated. Furthermore, this range also coincided with the range where there are convergence accommodation responses.
- The closer the average parallax distribution was to the screen, the more comfortable it was to watch the scene.

– Those scenes judged comfortable to watch showed a strong correlation between the sense of presence and the range of parallax distribution. The sense of presence increased as this range widened.
– No significant correlation was observed between the average of parallax distribution and the sense of presence.

In the future, we intend to study visual fatigue and visual comfort when viewing stereoscopic HDTV pictures that include scenes with more aggressive movement, scenes in which the parallax produces temporal changes, and long programs contain a variety of scenes. These results will help us to develop a stereoscopic HDTV system and techniques for producing attractive, fatigue-free stereoscopic HDTV programs.

4.3 Temporal Change in Parallax Distribution and Ease of Viewing

We have mentioned the difficulty of fusing right and left images as one cause of the visual discomfort of viewing stereoscopic images. A number of factors account for the right–left disparities, including binocular parallax [7, 8], geometric distortion [9, 10] (a vertical shift between right and left images, their size disparity), disparity in the brightness [10–12] and crosstalk [13]. Binocular parallax is inherent in stereoscopic images, but other factors are attributable to the performance and design of the system's equipment.

We conducted two new experiments to examine the relationship between visual comfort/discomfort and parallax distribution which are all important in stereoscopic HDTV program production. One experiment evaluated discontinuous temporal changes, and the other studied depth-wise position of superimposed characters in program production.

4.3.1 Discontinuous Temporal Changes in Parallax Distribution

In stereoscopic HDTV, it is sometimes necessary to directly connect two scenes with different parallax distributions during the editing process of program production. This procedure is called "cut change." This temporal change of parallax does not occur in the real world; it happens only in a stereoscopic HDTV program. To better understand how this affects the degree of visual comfort/discomfort, we conducted a subjective evaluation. First, we used geometric patterns with controlled parallax distribution to subjectively evaluate the visual discomfort caused by temporal changes in parallax. The results were then analyzed to find factors that contributed to this visual discomfort. Next, we examined the tolerance limit to this visual discomfort, which was regarded as impairment.

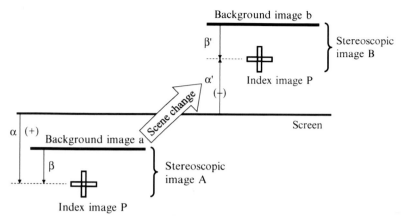

Fig. 4.20 Depth relationship of image used in evaluation

4.3.1.1 Factors Contributing to Visual Comfort/Discomfort

Subjective Evaluation Test

This section examines how, during the cut change, the visual discomfort of stereo-scopic images is affected by a change in the object's depth position and alters the viewer's eye convergence and the range of depth in the stereoscopic HDTV image. For the subjective evaluation test, we used two stereoscopic HDTV test images, A and B, that could be switched by cut change. As Fig. 4.20 shows, Image A and Image B each comprises a background image (a for A, b for B) and an index P. In this arrangement, the distance between index P and the screen is expressed by α for Image A and α' for Image B, and the distance between index P and background image by β for Image A and β' for Image B. The amount of change in the depth position of index P, expressed by $\Delta\alpha$, is given by:

$$\Delta\alpha = \alpha - \alpha'$$

$\Delta\beta$, which represents the change of depthwise width in the image, is given by:

$$\Delta\beta = \beta - \beta'$$

Random dot patterns were used for the background images a and b.

A cut-change pair of Image A and Image B, with a gray image in between, were shown twice to the subjects for the test. These images were evaluated on a scale of 5, as shown in Table 4.9, by the single-stimulus method. This evaluation term was based on the results of analyzing psychological factors related to the cut change. During the test, the viewers were asked to concentrate on the index image all the time. Table 4.10 shows the conditions of this test. The display device was a 90-inch projector with polarizing glasses. A total of 12 people, all non-experts, participated in the experiment.

Table 4.9 Scale of evaluation

2	Comfortable to watch
1	Somewhat comfortable to watch
0	Not sure
−1	Somewhat uncomfortable to watch
−2	Uncomfortable to watch

Table 4.10 Conditions of evaluation test

Screen size	90 inch (stereoscopic HDTV projector with polarizing glasses)
Viewing distance	Three times the screen height
Peak brightness	25 cd/m²
Method of evaluation	Single stimulus quality scale method
Subjects	12 non-experts

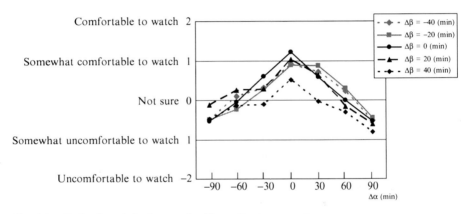

Fig. 4.21 Evaluation of viewing comfort/discomfort by scene change

Experimental Results

Figure 4.21 shows the evaluation results of visual comfort/discomfort, plotted by the sequential categorization method. Each plot represents the average of responses from the 12 test participants, who evaluated the test patterns twice. If $|\Delta\alpha| > 90$ min (an angle of minutes), Images A and B themselves might be difficult to watch comfortably. As the purpose of the experiment was to evaluate the visual discomfort as a result of cut change, we opted to show the results for up to 90 min of $|\Delta\alpha|$.

If $\Delta\beta$ is zero ($\beta = \beta' = 0$), the image whose index and background image are at the same depth position varies back and forth around the screen. Here, $\Delta\alpha$ is negative if the image moves toward the viewer and positive if it moves away from the viewer. It is clear from this graph that the image is uncomfortable to view when the depth-size change ($\Delta\alpha$) of the index image exceeds 60 min. Figure 4.21 also shows the results when the depthwise width ($\Delta\beta$) of the image was also changed. When $\Delta\beta = 40$ min, for instance, viewers began to experience visual discomfort past the 30 min of index position change.

4.3.1.2   Study of Contributing Factors by Multiple Regression Analysis

We employed a multiple regression formula to show the psychological scale S of visual comfort/discomfort, with the amount of depth-wise change $\Delta\alpha$ of the index and that of depth-wise width change $\Delta\beta$ of the image functioning as variables. The sign of $\Delta\alpha$ (i.e., whether the index is toward or away from the viewer) swerves as an explanatory variable g, and the sign of $\Delta\beta$ (i.e. whether the depth-wise width-wise increases or decreases) as explanatory variable h. Here, these variables are standardized as $\Delta\alpha^*$, $\Delta\beta^*$, g^*, and h^* such that their average becomes 0 and their dispersion 1. The standardized psychological scale S^* can be expressed by the following multiple regression formula:

$$S^* = 0.89 \, |\, \Delta\alpha^* \,| - 0.15 \, |\, \Delta\beta^* \,| + 0.06 \, g^* - 0.18h^*$$

where,

S^*: Standardized psychological scale of visual comfort
$\Delta\alpha^*$: Standardized depth-wise position change of index
$\Delta\beta^*$: Standardized depth-wise width change of the image
g^*: Standardized sign of $\Delta\alpha^*$
h^*: Standardized sign of $\Delta\beta^*$
A comparison of their coefficients reveals that:

- The coefficient of $\Delta\alpha^*$, −0.89, is larger than the others. This indicates that, when the image's parallax distribution is altered, a change in the depth-wise position of the watching point largely influences visual comfort.
- The coefficient of g^*, 0.06, is smaller than the others (it became zero at the significance level of 5%). This means that there is no significant difference whether the watching point moves toward or away from the viewer.
- The coefficient of $\Delta\beta^*$ and that of h^* are −0.15 and −0.18, respectively. These figures indicate that any change in the image's depth-wise width will have less impact on the visual comfort than changes in the watching point's depth-wise position, and that viewers will begin to feel slightly more uncomfortable as the depth-wise width increases.

4.3.1.3   Tolerance Limit of Depth Change

These results indicate that, when the image's parallax distribution is altered by a cut change, visual comfort is largely influenced by convergence-shifting changes in the depth-wise position of the watching point. We can therefore set the tolerance limit of this depth-wise position change by regarding it as an impairment. Assuring that $\beta = 0$, we changed only α as we conducted an experiment to subjectively evaluate the test images on the five-point impairment scale [14]. A total of 15 people participated in the experiment under the same conditions as those shown in Table 4.10.

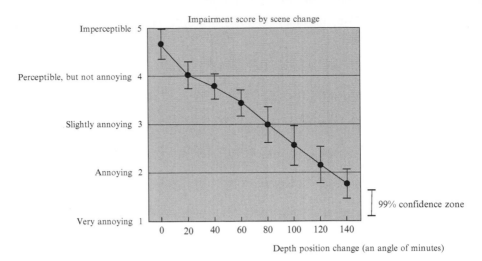

Fig. 4.22 Evaluation of impairment by changes in depth position

Figure 4.22 shows the results. Each plot represents the average of responses from the 15 viewers who observed the test images twice. The error bar represents a 99% confidence interval. The tolerance limit of the watching point's depth-wise position change, the point where the evaluation value is 3.5 in the graph, is about 60 min.

4.3.2   Superimposed Characters

We also conducted an evaluation test on visual comfort/discomfort, focusing on the depth position of superimposed characters that play an important role in stereo-scopic HDTV programs. Figure 4.23 shows four pictures, each with superimposed characters at a different depth position. The conditions of the experiment, conducted by the adjustment method, are shown in Table 4.11. In this experiment, subjects watched stereoscopic images like those in Fig. 4.23 while freely moving the depth position of the superimposed characters with a track ball. Each subject was asked to determine the point at which the image, including the superimposed characters, was the most comfortable to view. The results are shown in Fig. 4.24. The names on the horizontal axis represent the images used in this subjective evaluation. Table 4.12 shows some characteristics of the parallax of the background where the superim-posed characters are located. Picture patterns are the same among Flower 1, 2, 3 and Shrine 1, 2, 3, but their parallax distribution differs because of different shooting conditions. The numbers on the vertical axis represent the differences in conver-gence angle between when the viewer's attention is on the superimposed characters and when it is on a point of the background image closest to him. Each dot in Fig. 4.24 represents the average of evaluation points given by the 22 test participants

1. City   2. Room

3. Flower   4. Shrine

Fig. 4.23 Images used in the evaluation test

Table 4.11 Test conditions

Screen size	90 inch (stereoscopic HDTV projector with polarizing glasses)
Viewing distance	Three times the screen height
Peak brightness	30 cd/m^2
Method of evaluation	Adjustment method
Subjects	22 non-experts

Fig. 4.24 Optimal depth position of superimposed characters

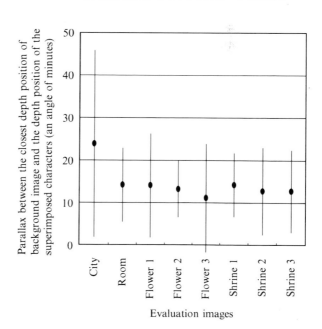

Table 4.12 Characteristics of parallax in evaluation images

Evaluated images	Convergence angle (min) at closest background image	A range of parallax distribution (min) of background image
City	38	48
Room	53	10
Flower 1	72	28
Flower 2	78	40
Flower 3	120	60
Shrine 1	72	16
Shrine 2	84	31
Shrine 3	84	50

Table 4.13 Influence of parallax distribution on visual comfort/discomfort

Parallax distribution	Experimental results	Note
Parallax distribution in an image (depth-wise width)	Comfortable to watch if parallax distribution is within 60 min	Preferred depth position of superimposed characters in stereoscopic HDTV image: 10–15 min toward the viewer
Temporal change in parallax distribution (temporal change of depth)	Uncomfortable to watch if the temporal change of parallax exceeds 60 min	Discontinuous temporal change in parallax distribution (assuming cut change)

and the vertical bar represents standard deviation. This graph suggests that the superimposed characters are the most comfortable to see when they are about 10–15 min in parallax toward the viewer from the closest point of the background. An exception is "City", where the comfortable-to-view position of the superimposed characters varied widely from one viewer to another.

These results show that the depth position of superimposed information must be projected toward the viewer from the background, but closeness to the viewer varies depends on personal preference. Generally, viewers seem to prefer the point about 10–15 min toward them.

4.3.3 Summary

Table 4.13, which shows how the visual comfort is affected by parallax distribution, shows the results of our subjective evaluation test, in which test participants watched stereoscopic HDTV pictures under standard viewing conditions for conventional HDTV. Note that both the range of parallax distribution and the amount of discontinuous temporal changes in the parallax distribution with which viewers found it comfortable to watch stereoscopic HDTV images were about 60 min [15]. Here this range of parallax distribution is defined as the difference between two convergence angles on opposite ends—one is when the viewer's convergence point is on an object projected nearest toward him and the other is when the convergence point is on an object farthest away from him.

When producing stereoscopic HDTV programs, it is important to pay close attention to the range of parallax distribution and its changes during shooting and editing, and determine to what extent the program director wants stereoscopic effects.

References

1. H. Yamanoue, M. Emoto, M. Okui, S. Yano and T. Yoshida, "Stereoscopic Test Materials", Proc. IDW 3-D3-2, pp. 815-818, 1998
2. H. Yamanoue and F Okano, "Hi-vision Stereoscopic Harp Camera", NHK R&D, Vol.31, pp. 58-67, 1994
3. G.A. Thomas, "Television Motion Measurement for Other Applications", BBC Research Department Report, Sept., 1987
4. M. Okui, H. Yamanoue, M. Emoto, S. Yano and T. Yoshida, "Test materials for evaluating stereoscopic television systems", Proc. IBC, pp. 565-570, 1999
5. W. N. Charman and H. Whitefoot, "Pupil Diameter and the Depth-of-field of the Human Eye as Measured by Laser Speckle", OPTCA ACTA, 24, 12, pp. 1211-1216, 1977
6. N. Hiruma and T. Fukuda, "Accommodation Response to Binocular Stereoscopic TV Images and Their Viewing Conditions", SMPTE J., Vol.102, pp. 1137-1144, 1993
7. K.N. Ogle, "On the Limits of Stereoscopic Vision", J. Experimental Psychology, 44, pp. 253-259, 1952
8. M. Wopking, "Viewing comfort with stereoscopic pictures: An experimental study on the subjective effects of disparity magnitude and depth of focus", Journal of the SID, Vol.3, pp. 101-103, 1995
9. F. Speranza and L. M. Wilcox, "Viewing stereoscopic images comfortably: the effects of whole-field vertical disparity", Proc. SPIE, Vol.4660, pp. 18-25, 2002
10. B. Choquet, "3-DTV studies at CCETT", TAO 1st International Symposium 1993
11. I. P. Beldie and B. Kost, "Luminance asymmetry in stereo TV images", Proc. SPIE, Vol.1457, pp. 242-247, 1991
12. J. Fournier and T. Alpert "Human factor requirements for a stereoscopic television service: Admissible contrast differences between the two channels of a stereoscopic cameras", Proc. SPIE, Vol.2177, pp. 45-54, 1994
13. A. Hanazato, M. Okui and I. Yuyama, "Subjective Evaluation of Cross Talk Disturbance in Stereoscopic Displays", Proc. SID, pp. 288-291, 2000
14. "METHODOLOGY FOR THE SUBJECTIVE ASSESSMENT OF THE QUALITY OF TELEVISION PICTURES", Rec. ITU-R BT. 500-10
15. Y. Nojiri, H. Yamanoue, A. Hanazato and F. Okano, "Measurement of parallax distribution, and its application to the analysis of visual comfort for stereoscopic HDTV", Proc. SPIE, Vol.5006, pp. 195-205, 2003

Chapter 5
Visual Fatigue When Viewing Stereoscopic Television with Binocular Parallax

Abstract Stereoscopic 3-D television broadcasting services are expected to underpin the imaging environments of the future, but several problems must still be solved before 3-D television can enter widespread, popular use. Visual fatigue stands out as the most important of these. With present systems, an audience of several hundred at the 3-D theater of an amusement park, for instance, complains of visual fatigue after watching 3-D images for only 10–20 min. If the same were to happen after only a couple of hours with a 3-D TV broadcasting service, the consequences and very feasibility of the service would necessarily come into question. Even if the effects on human health are not serious enough to require medical assistance, no broadcasting service that causes many viewers to suffer from visual fatigue could be accepted. It is, therefore, essential to devise a system prior to implementation that enables a large number of viewers with differing characteristics to watch 3-D TV under various viewing conditions for long periods without apparent visual fatigue. Unlike viewers of conventional TV, the viewers of 3-D TV make full use of binocular stereopsis function and are, therefore, liable to experience visual fatigue. Here, we examine characteristics of visual fatigue due to binocular stereopsis and consider possible solutions. Issues for the Evaluation and Elimination of Fatigue.

The key challenges for evaluating fatigue are:

1. The subjectivity of fatigue, which cannot be measured directly. It is necessary to establish an objective index.
2. The broadness of the term, which allows considerable diversity and ambiguity of interpretation. This diversity and ambiguity has to be excluded.
3. Identifiable causes of visual fatigue, such as visual dysfunction and disease, can be reduced or eliminated by the ophthalmologist but viewers often complain of visual fatigue for no identifiable reason.

H. Yamanoue et al., *Stereoscopic HDTV: Research at NHK Science and Technology Research Laboratories*, Signals and Communication Technology, DOI 10.1007/978-4-431-54023-6_5, © Springer Japan 2012

The following solutions are proposed in this chapter:

1. Fatigue studies in labor and other sectors have been based on the view that any change in bodily functions under load reflects fatigue. They measure functional deterioration, so deterioration of visual function becomes the marker of visual fatigue. In this study, we use the state of visual functioning observed directly before and after viewing binocular parallax-based 3-D TV as our objective index.
2. This study considers only visual fatigue, not fatigue in general.
3. If visual fatigue during and after viewing 3-D TV is due to visual malfunction or disease, it is regarded as a matter for medical treatment and is disregarded here. We also disregard fatigue due to indefinable causes, limiting our study of visual fatigue in this chapter to fatigue that has identifiable causes.

Keywords 3-D • AC/A ratio • Accommodation step response • Area of comfort • Binocular parallax • Fusion range • P100 • Range of relative vergence • Stereoscopic TV • Vergence • Visual evoked potential (VEP) • Visual fatigue

5.1   Establishing an Objective Index for Evaluating Visual Fatigue

It is essential to establish a quantitative index of objective numbers for use in fatigue studies in tandem with subjective evaluations. However, no such quantitative index has been established for use in assessing the fatigue of watching binocular parallax-based stereoscopic television. Visual fatigue such as eyestrain has been evaluated subjectively (in the 3-D theater of an amusement park) but without quantitative comparison of fatigue levels or identification of what causes visual fatigue. Establishment of such an index to evaluate visual fatigue objectively is a vital first step in assessing fatigue experienced when viewing binocular parallax-based stereoscopic television. Some of this fatigue may be caused by the same factors as in conventional flat-screen 2-D television, namely rapid motion, frequent scene changes, and inadequate resolution on the large screen; in this study we address only the causes of visual fatigue that are specific to stereopsis.

We focused here on two kinds of visual function among the stereopsis functions related to binocular vision, namely the range of relative vergence (because the ocular vergence function is particularly overused when viewing binocular parallax-based stereoscopic television) and the AC/A ratio (the ratio of accommodative convergence to accommodation; it shows the strength of the interaction between accommodation function and vergence function) because this interaction works differently from stereopsis in the real world. The range of relative vergence is defined as the difference in the fusional vergence break points between convergence and divergence, and it is the viewer's tolerance range for varying ocular vergence with almost no change in accommodation. We hypothesized that assessment of these functions before and after viewing could be used as a measure of visual fatigue.

We applied a fatigue load using parallax-based stereoscopic 3-D television and examined the effects on fatigue after viewing and after rest. The effects of 2-D television on fatigue were used as a control condition to compare with 3-D. We also investigated the difference in the effects between test subjects who were able to freefuse and who were unable. It is extremely important to study the relationship between rest and recovery in our evaluation of visual fatigue. If the visual fatigue resulting from viewing 3-D TV is reduced by rest and does not become a chronic, pathological form of visual fatigue, then 3-D TV broadcasting may indeed be feasible. Conversely, if visual fatigue has serious effects on visual functioning, especially on the visual functioning of developing children, 3-D TV broadcasting would be out of the question.

5.1.1   Subjective Evaluation

Figure 5.1 shows the results of a questionnaire on the subjective feelings of fatigue experienced after viewing television for an hour.

Five test subjects found the 3-D images are more tiring than 2-D images; six found them equally tiring; and one found the 2-D images more tiring than 3-D images. These responses suggest that 3-D images are significantly more tiring to view than those in 2-D.

5.1.2   Results for AC/A Ratio

Figure 5.2 shows average changes for stimulus AC/A ratio, as measured by the gradient method before and after television viewing. The error bars indicate standard deviation. Analysis of variance found that the main effects of time ($F(1, 8)=0.620$, $p=0.454$), the freefuse ability ($F(1, 8)=2.16$, $p=0.180$), and interaction ($F(1, 8)=0.620$, $p=0.454$) were not significantly different when viewing 3-D images. In the case of TV watching for only an hour, therefore, we did not detect any adaptation (such as reduction of accommodative convergence or substantive change in the AC/A ratio) as a result of visual fatigue, suggesting that the interaction between accommodation and convergence did not change significantly.

5.1.3   Results for the Range of Relative Vergence

Taking the range of relative vergence as "1" before viewing television, we sought the normalized range immediately after viewing and again after some rest. Figure 5.3 shows the averaged changes.

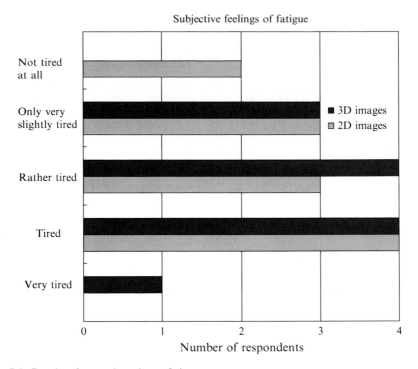

Fig. 5.1 Results of a questionnaire on fatigue

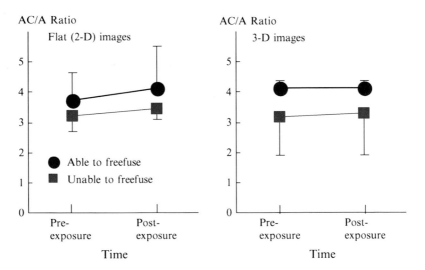

Fig. 5.2 Changes in AC/A ratio

Fig. 5.3 Changes in
normalized ranges of relative
vergence

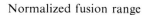

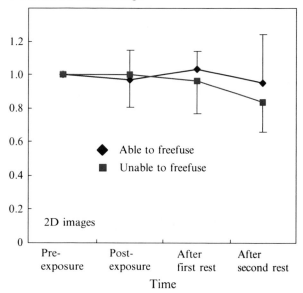

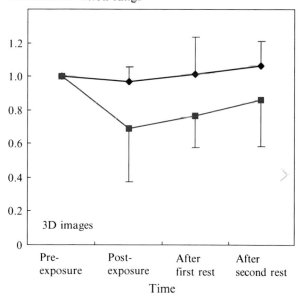

Regarding changes in the normalized range of relative vergence when viewing
3-D images, we could assume sphericity on the basis of Maunchly's sphericity test
($p = 0.787$). Repeated measures analysis of variance found that the main effects
of freefuse ability tended to be significant ($F(1, 10) = 4.62$, $p = 0.057$). Further,

we observed main effects of time in the range of relative vergence ($F(3, 30) = 3.58$, $p = 0.025$) and these interactions were significant ($F(3, 30) = 2.55$, $p = 0.075$).

For changes in the range of relative vergence when viewing 2-D images, we conducted repeated measures analysis of variance in the same way as when viewing 3-D images, and found that both the main effects and interactions were insignificant.

The fact that the interactions were significant with regard to changes in the normalized range of relative vergence when viewing 3-D images suggested the tendency of change in the range of relative vergence might differ between subjects who are able to freefuse and those who are unable. It also seemed that visual load and the tolerance of fatigue may vary when viewing 3-D images. We performed Dunnett's multiple comparisons separately by freefuse ability and found no significant difference between the situation before viewing 3-D images and at the other times in test subjects with freefuse ability.

With regard to test subjects without freefuse ability, significant differences were found between before viewing 3-D images and the other times ($p = 0.017$). Although not significant, a tendency was found after the first rest ($p = 0.079$), no significant difference was found after the second rest ($p = 0.394$). This indicated that the test subjects had almost completely recovered from visual fatigue after that second rest. It also suggested that changes in the range of relative vergence are different for test subject freefuse ability.

A previous study found changes in the range of relative vergence of visual display terminal (VDT) operators who had been working for 6 h [1]. Differences in viewing distance for 3-D image viewing and work with VDTs, together with the different concentration and attention levels required for such work and differences in convergence and accommodation, make it impossible to compare these results directly, but this also suggests that the range of relative vergence does indeed reflect visual fatigue.

Whereas the range of relative vergence scarcely changes when viewing 2-D images, it becomes significantly narrower immediately after 3-D image viewing than either before viewing or after a rest, when it returns to almost the same level as prior to viewing. This change of range of relative vergence is thought to reflect visual fatigue caused by image viewing and the recovery process. We conclude that changes in the range of relative vergence may be regarded as an effective index for the objective evaluation of visual fatigue produced when viewing 3-D images.

5.1.4   Summary

The visual fatigue of watching 3-D images has been reported subjectively in the past, and we did corroborate those subjective evaluations in our experiment. It is also clear, however, from the objective changes in the range of relative vergence before and after viewing that it is more tiring to watch 3-D than 2-D images. These results imply that visual functions associated with binocular fusion are strongly implicated in the visual fatigue experienced when viewing binocular parallax-based stereoscopic television.

When observers see in the real world, the convergence and divergence of the eyes that occur when the eye lines need to be adjusted for stereoscopic fusion are involuntary. No such motion occurs if no attention is paid to observing objects, but it cannot be willfully suppressed when an individual needs to focus on objects. The double image of an object to which attention is not directed in the perception of stereoscopy is neither a particular problem nor a cause of visual fatigue. If the object to which attention is directed is perceived as a double image, however, this can produce strong discomfort and has been associated with eyestrain. This eyestrain is less significant when the double images are too far apart to fuse, but eyestrain becomes intense if the overlap is close and the images can be fused with strong effort. A double image that cannot be fused by any means stimulates no further effort and no eyestrain results. Accordingly, eyestrain arises with double images within certain limits.

When images with large binocular parallax are shown on 3-D TV, the viewer makes a considerable effort to fuse the left and right images; this is very likely to produce visual fatigue. Binocular parallax-based 3-D TV can present fusion-difficult left and right images to the viewers as with a real object; furthermore, it can present more fusion-difficult left and right images than the real object, depending on shooting and viewing conditions. This effect might cause a great deal of fusion effort in attempts to fuse two retinal images which can never exist in the real world.

Thus, we propose a hypothesis that the visual fatigue experienced when viewing parallax-based 3-D TV may be chiefly due to the effort expended on fusing the fusion-difficult left and right images, as in the case of the visual fatigue inherent to stereopsis in the real world. This hypothesis is consistent with the accepted plausible cause of visual fatigue, which is the inconsistency of binocular convergence and accommodation when watching parallax-based 3-D TV [2]. It is easy to understand why this inconsistency produces visual fatigue, as more effort has to be expended on fusion as the inconsistency of binocular convergence and accommodation becomes larger.

5.2   Two Factors of Visual Fatigue

The difference between left and right images, which makes fusion effort necessary, can be understood in two ways in the case of the binocular parallax-based 3-D TV system. One factor is attributable to horizontal binocular parallax, providing a direct clue for depth; this is unavoidable in parallax-based 3-D TV and is termed the "in-principle" factor. The other factor consists of the difference between left and right images attributable to differences between the left and right characteristics of the imaging equipment, and is essentially unnecessary in binocular parallax-based 3-D TV; this is termed the "non-principle" factor. In the experiment above, we studied overall visual fatigue resulting from viewing parallax-based 3-D TV with both factors mixed.

Fundamental examination is needed to build a 3-D TV system that minimizes visual fatigue. Even though some differences between the left and right images attributable to the differences between the left and right characteristics of imaging equipment can be removed by technological improvements, the essential inevitability of visual fatigue beyond the tolerated limits defies technological solution. We shall, therefore, simply examine the effects on visual fatigue of the in-principle factor first, to verify the hypothesis that visual fatigue when viewing binocular parallax-based 3-D TV is due to the effort required to fuse the left and right images. In the next section, we investigate visual fatigue by only the in-principle factor using an optical simulator of 3-D TV viewing. We refer to 3-D TV with only the in-principle factor as "ideal binocular parallax-based stereoscopic TV" in which the non-principle factor is removed.

5.3   Evaluation of Visual Fatigue When Viewing Ideal Binocular Parallax-Based Stereoscopic Television

Introduction

We controlled the vergence load burden for each test subject in order to simulate an ideal binocular parallax-based stereoscopic television in which only the binocular parallax, the in-principle factor of binocular parallax-based 3-D TV, exists, and varied the amount of effort required for fusion. We used prisms to produce the effect of watching images on the ideal 3-D TV.

The visual functions work as shown in Fig. 5.4 when viewing 3-D TV. When the object is perceived in front of the screen, the vergence point is located in front of the depth of field and this produces excessive convergence. When the object is perceived to be behind the screen, the depth of field is located in front of the vergence point and produces excessive eye accommodation.

Figure 5.5 shows a simulation of this viewing situation where convergence is induced by prism load. If the prisms are placed with base facing outwards (base out: BO) before the test subject's eyes, each producing a single view of the single image, the two eyes will be adducted (converge) to correct displacement of the light by the prism and retain a single, fused image. Accommodation will be maintained so that the image does not blur. This is equivalent to the excessive convergence experienced when viewing 3-D images as in Fig. 5.4 front. Conversely, if the prism is placed with its base facing inwards (base in: BI) in front of the eyes, they will be abducted (diverge). This is equivalent to the excessive accommodation experienced when viewing 3-D images as in Fig. 5.4 rear. It is possible in this way to place a certain load on the vergence system only and examine visual fatigue according to this load.

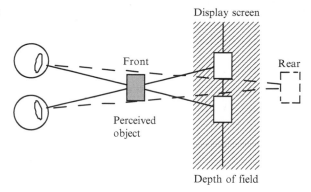

Fig. 5.4 Relationship between convergence and accommodation when viewing 3-D images

Display screen

Front

Rear

Perceived object

Depth of field

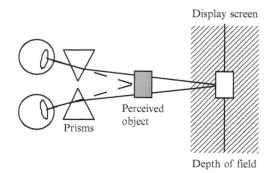

Fig. 5.5 A simulation of 3-D image viewing using prisms

Display screen

Perceived object

Prisms

Depth of field

5.3.1   Evaluating Visual Fatigue by Prism Load

In the experiment of visual fatigue evaluations by prism load, each subject viewed images for about an hour under the conditions illustrated in Fig. 5.5. We gathered subjective reports and performed objective evaluations according to changes in visual function. Subjective evaluations were made in terms of eyestrain on a scale of 1–5, where 1 = very much and 5 = not at all.

As objective indices, we used three items relating to eye convergence and divergence: the range of relative vergence found to be effective as an index for evaluating visual fatigue in the previous study (see 5.1), accommodation step response, and visual evoked potentials (VEPs) apparently connected with the central nervous system.

In visual fatigue research, individual differences often cause problems. Even if the stimulus is the same, the load can vary due to the diversity of individual visual functions. Looking at the area of comfort (Note 1), we determined the amount of vergence load for each test subject as a ratio of their range of relative vergence (defined between breakpoints in convergence and divergence), and attempted to establish experimental conditions that would allow for those individual differences.

Table 5.1 Breakpoints in convergence and divergence

	Convergence side (BO) limit (Δ)	Divergence side (BI) limit (Δ)
Test subject 1	−12	30
Test subject 2	−10	6
Test subject 3	−10	12
Test subject 4	−12	28
Test subject 5	−12	28
Test subject 6	−10	20

We also studied what effects individual differences in the range of relative vergence might produce on visual fatigue.

Note 1

Area of comfort: Percival further developed Landolt's theory that the subject might not be able to use more than one-third of his or her absolute convergence continually without visual fatigue, and defined an area of comfort. This area of comfort occupies the middle third of the total range of relative vergence and the angle of convergence is limited to the range of infinity to 33 cm [3].

5.3.2   Test Conditions

To begin, we measured the convergence and divergence breakpoints for each test subject to determine their area of comfort. Base-in and base-out limits were determined to the closest integer by rounding the measured breakpoints (omitting the figures below the decimal place to keep the middle third of the relative range within the area of comfort). Each subject's BO and BI limits are shown in Table 5.1. We determined the amount of vergence load for each test condition on the basis of each subject's area of comfort as obtained from the limit values. We set up six static conditions to place a certain load on the vergence system and four dynamic conditions to place load on temporal change.

The static conditions with no change in depth were:

1. Center condition; the prism power was at the center of the ranges of relative vergence (center of Percival's area of comfort).
2. "−1/3" condition; the power of the BI prisms was within the middle third of the range of relative vergence (divergence side of the "area of comfort").
3. "+1/3" condition; the power of the BO prisms was within the middle third of the range of relative vergence (convergence side of the "area of comfort").
4. "−2/3" condition; the power of the BI prisms was the average of −1/3 condition and BI limit (divergence side of "out of the area of comfort").
5. "+2/3" condition; the power of the BO prisms was the average of +1/3 condition and BI limit (convergence side of "out of the area of comfort").

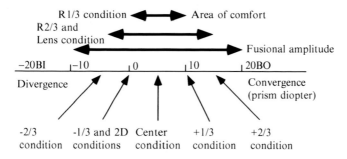

Fig. 5.6 Test conditions

Figure 5.6 shows the relationships between these test conditions and the amount of parallax for a subject with a fusional BO limit of $+20\Delta$ and fusional BI limit of -10Δ.

6. "2-D" condition; when the prism power calculation was between -2Δ and $+2\Delta$. In the above conditions, the prism power was determined to be 0Δ. This was used as the 2-D condition for data processing.

Six static conditions were established in this way. There were test subjects who could not fuse presented images from the very beginning in the +2/3 condition and in those cases; we reduced the prism power until fusion occurred. It is possible to fuse comparatively large binocular parallax when a continuous increase of parallax produces the limit value, but the suddenness makes such fusion impossible in the +2/3 condition. Under the above test conditions and for a normal ocular position, the amount of inconsistency in convergence and accommodation increases as the depth of the scene moves away from the middle of Percival's area of comfort (the center condition).

Temporal changes due to the following dynamic conditions are necessary to achieve convergence in conjunction with changes of scene depth:

7. "Random 1/3" (R1/3) condition; random selection of prism power within Percival's area of comfort for both BI and BO.
8. "Random 2/3" (R2/3) condition; random selection of prism power within the middle two-thirds of the total range of relative vergence outside Percival's area of comfort.
9. "Lens" condition; the addition of lens compensation for accommodation at the induced convergence point to the random selection of the prism power within the middle two-thirds of the total range of relative vergence outside Percival's area of comfort.
10. "3-D" condition; presentation of a general stereoscopic image with binocular parallax to the left and right eyes without prisms using dichoptic displays.

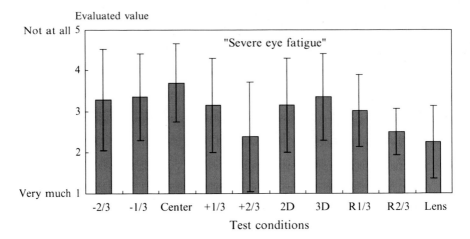

Fig. 5.7 Subjective evaluations of severe eye fatigue

5.3.3   Results

5.3.3.1   Subjective Evaluations

After viewing the 3-D images for about an hour, subjective symptoms of eyestrain were assessed on a five-point scale. Responses by the test subjects were quantified in the five categories from "Very much" = 1 to "Not at all" = 5. Figure 5.7 shows the average results for "severe eye fatigue" for subjects under various test conditions. The responses were so varied by test subject that we did not detect any statistically significant differences between the test conditions. The scores were best, however, in the center condition and tended to decline as convergence and accommodation became increasingly inconsistent under dynamic conditions. As for the difference between the convergence and divergence sides, the scores were better for the convergence side in general.

5.3.3.2   Range of Relative Vergence Ratios

Figure 5.8 shows range of relative vergence ratio averages for test subjects under different conditions before and after viewing the images. The pre- and post-viewing values were compared by paired t-test. The results showed significant differences in the $-2/3$ ($p=0.0307$, $t(5)=2.40$), $-1/3$ ($p=0.0301$, $t(5)=2.42$), $+2/3$ ($p=0.0300$, $t(5)=2.42$), 3-D ($p=0.0126$, $t(5)=3.15$) and R2/3 ($p=0.0216$, $t(4)=2.92$) conditions (asterisks indicate significant difference).

When the vergence load was either constantly heavy or changed substantially over time, the range of relative vergence decreased significantly after image viewing. This is taken to reflect the visual fatigue felt after viewing under vergence load.

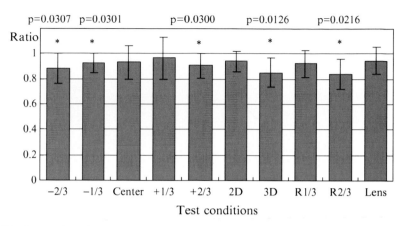

Fig. 5.8 Range of relative vergence ratios before and after viewing images

With lens compensation for accommodation, there was no significant change in the range of relative vergence ($p=0.160$, $t(4)=1.13$).

Under the five test conditions, the range of relative vergence decreased significantly right after image viewing, this might reflect visual fatigue. As discussed earlier, it was confirmed that subjects who were able to freefuse had a wider range of relative vergence than those who were unable, and the variation in the range of relative vergence when watching 3-D images was also different in this second group. In this experiment, we supposed that the variation may have been due to different competence levels in convergence and divergence, and therefore studied the differences in those terms. We grouped the subjects into those who had a wide range of relative vergence (Subjects 1, 4, and 5 in Table 5.1) and those with a narrow range of relative vergence (Subjects 2, 3, and 6 in Table 5.1), then examined their convergence and divergence abilities and the interactions over time. Multiple measurements were needed to investigate interactions, so we made further measurements after allowing a rest period. Specifically, we performed repeated measures analysis of variance with four time points: before image viewing, after image viewing, after a first rest period, and then after a second rest period. Figure 5.9 shows the normalized average variation in the range of relative vergence for the six subjects, taking the range of relative vergence as 1 before image viewing. The result from this analysis showed the validity of Mauchly's sphericity assumption under all of these conditions. Significant differences were detected for the main effects of time in the −2/3 ($p=0.014$), −1/3 ($p=0.0012$), +2/3 ($p=0.003$), 3-D ($p=0.0141$), R1/3 ($p=0.00496$), and R2/3 ($p=0.00556$) conditions. The interaction was significant in the +2/3 condition ($p=0.040$), implying that the changes in the range of relative vergence in the +2/3 condition were related to the subject's vergence capabilities.

Next, Dunnett's multiple comparison was performed to investigate the −2/3, −1/3, 3-D, R1/3 and R2/3 conditions, in which significant differences had been found for the main effects of time but not for interactions. Significant differences were detected immediately after image viewing in the −2/3 ($p=0.020$), −1/3

Range of relative vergence ratios

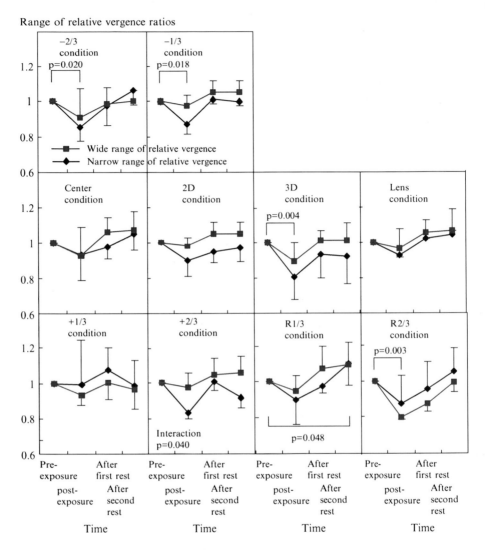

Fig. 5.9 Range of relative vergence ratios

(p=0.018), 3-D (p=0.004) and R2/3 (p=0.003) conditions. These results corresponded to those of the paired t-test. Except in the R1/3 condition, no significant difference was found after image viewing, suggesting that the test subjects had almost recovered from visual fatigue during their rest.

It was not immediately after image viewing (p=0.136), but only after the second rest (p=0.048), that significant differences were found in the R1/3 condition. The range of relative vergence then became significantly broader than before image viewing. It is suggested that a moderate repetition of eye movement within Percival's area of comfort increases the range of relative vergence.

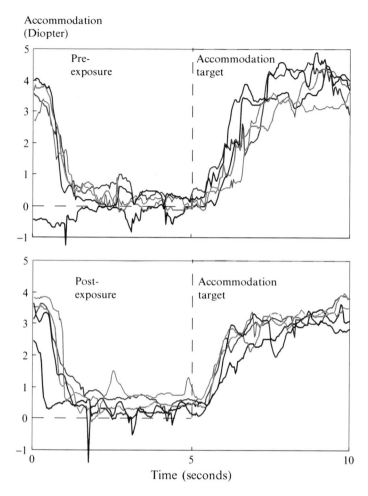

Fig. 5.10 Example of varied accommodation response

5.3.3.3 Accommodation Step Response

The test subject's accommodation response on the optical position index (accommodation target), capable of producing stepped changes from 0 to 5 diopters, was measured five times. The comparison of response waveforms before and after image viewing did not reveal any systematic change common to all subjects. Under several test conditions, however, some of the subjects did show variation in their near viewing accommodation responses after image viewing. Figure 5.10 shows the example of a subject with varied accommodation step response, while Fig. 5.11 gives an example of unvaried stepped accommodation response. In the following figures, the dotted line at 5 s from 0 to 5 diopters shows changes in accommodation target depth, while the five unbroken lines show the results of the subject's accommodation responses measured five times.

Accommodation
(Diopter)

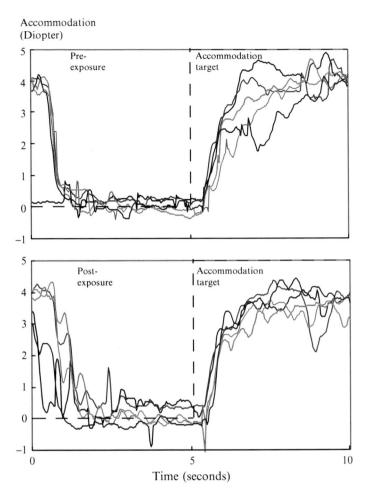

Fig. 5.11 Example of unvaried accommodation response

5.3.3.4   P100 Latency of Visual Evoked Potentials

Visual evoked potentials (VEPs) are changes in brain potential induced by visual stimuli, such as flashing lights or pattern reversal. The VEP is measured by electrodes on the scalp when impulses photoelectrically converted in the retina reach the visual cortex of the occipital lobe via the optic nerve and lateral geniculate nuclei. This electrical potential is very small (as microV), so the visual stimulation is repeated and the obtained potentials are averaged from the starting point of stimulation to cancel recorded noise. A polyphasic waveform is obtained when the repetition frequency of the visual stimulation is as low as several Hz; this is known as the transient VEP. This waveform makes it possible to measure the time to each peak (latency) and consider the changes in latency before and after

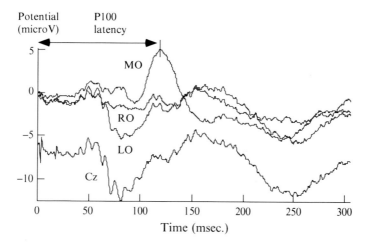

Fig. 5.12 VEP waveforms

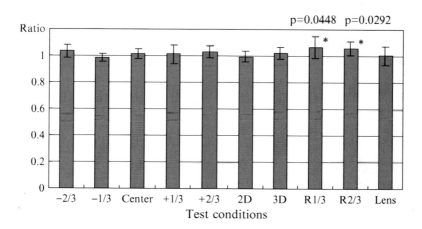

Fig. 5.13 Ratio of P100 latency before and after image viewing

image viewing. Figure 5.12 shows example VEP waveforms and P100 latency. In this experiment, we gauged the latency of the P100 component in the mid-occipital (MO) region.

Figure 5.13 shows the averaged results, by test condition, of subjects in terms of the ratio of the latency of P100 after image viewing. The paired t-test did not find any significant change under static conditions. In dynamic conditions, significant delays were found in the R1/3 ($p=0.0448$, $t(5)=-2.10$) and R2/3 ($p=0.0292$, $t(4)=-2.63$) conditions. No significant change was found with lens compensation for inconsistent convergence and accommodation ($p=0.453$, $t(4)=-0.127$).

5.3.4   Summary

We organized hour-long image viewing with prismatic simulation of an ideal stereo-scopic 3-D TV and assessed the changes in range of relative vergence, accommodation step response, and visual evoked potentials before and after image viewing, reaching the following conclusions:

The visual fatigue experienced when viewing 3-D images is attributable to large binocular parallax that exceeds Percival's area of comfort. Visual fatigue is also attributable to the discontinuous temporal changes in binocular parallax, due to changes of depth and scene changes in the images being viewed. This is partly because large binocular parallax close to the limits of fusion requires fusion effort, which is a visual load, and convergence without accommodation demand can cause visual fatigue.

The second reason for visual fatigue is that repeated adaptation to temporal change in the relationship between convergence and accommodation imposes a heavy burden. It is, therefore, preferable to restrict the amount of binocular parallax and avoid discontinuous temporal changes in the binocular parallax in order to reduce visual fatigue. By restricting binocular parallax, however, we also fear that the attraction of 3-D images will be diminished.

The challenge remains for 3-D content creators to determine the acceptable range for such parallax restriction while still making attractive 3-D content with sufficient depth perception.

References

1. B. Piccoli, A. Zaniboni, M. Meroni and A. Grieco: "Change in visual function and viewing distance during work with VDTs," Ergonomics, 33, 12, pp. 1433-1441 (1990).
2. T. Inoue and H. Ohzu, "Accommodative responses to stereoscopic three dimensional display," Appl. Opt., vol. 36, no. 19, pp. 4509-4515 (1997).
3. Percival, A.S.: "The relation of convergence to accommodation and its practical bearing," Ophthal. Rev. 11, pp. 313-328 (1892).

Index

H. Yamanoue et al., *Stereoscopic HDTV: Research at NHK Science and Technology* 125
Research Laboratories, Signals and Communication Technology,
DOI 10.1007/978-4-431-54023-6, © Springer Japan 2012

Printed by Publishers' Graphics LLC
SO20120714